The River Chief System and An Ecological Initiative for Public Participation in China

Yaguang Hao · Tingting Wan

The River Chief System and An Ecological Initiative for Public Participation in China

Springer

Yaguang Hao
Institute for China Rural Studies
Institute for Advanced Study in Political
Wuhan, China

Tingting Wan
Institute for China Rural Studies
Institute for Advanced Study in Political
Wuhan, China

ISBN 978-981-99-4920-5 ISBN 978-981-99-4921-2 (eBook)
https://doi.org/10.1007/978-981-99-4921-2

This Springer imprint is published by the registered company Springer Nature Singapore Pte Ltd.
The registered company address is: 152 Beach Road, #21-01/04 Gateway East, Singapore 189721, Singapore

Contents

Chapter 1
Introduction: River Chief System and Ecological Progress in China

1 Background

River Chief System is an innovative policy initiative to tackle water pollution in Taihu Lake in Jiangsu and Zhejiang provinces in 2007. River Chiefs are the key leaders of the Chinese Communist Party and governmental officials at provincial, municipal, county and town levels, who are responsible for combatting river pollution and protecting and managing rivers within their jurisdiction. Under the River Chief System, the Party and governmental leadership responsibility system is established. Not only do River Chiefs manage and coordinate inter- and intra- jurisdictional water protection and related affairs, but also, they shoulder the work of supervising river chiefs at lower levels in their jurisdiction. In December 2016, the General Offices of Central Committee of the Chinese Communist Party and the State Council jointly issued the "Opinions on promoting the River Chief system" to promulgate the River Chief System nationwide to prevent river pollution and protect water ecology in China. Since the General Office of the CPC Central Committee and the General Office of the State Council issued the "Opinions on the Comprehensive Implementation of the River Chief System" in 2016, China's River Chief System has undergone a profound transformation, from the establishment of institution and responsibility to the introduction of strict control to improve the water quality of rivers and lakes. This has created a new situation of water control and rejuvenation of water governance in the new era.

In the past five years since the full implementation of the river chief system, water pollution in China has been effectively curbed, black and odorous waters in cities at the prefecture level and above have been substantially eliminated, and the proportion of Class I to III water quality sections has increased by nearly 16 percentage points nationwide in 2020 compared with 2016. Zhejiang Province, for example, has intensively implemented the "Five Water Co-Governance", carried out various special corrective measures such as the "purification of three types of rivers" and the "elimination of low-quality class V water", and further improved the ecological environment of water (Tian et al., 2022). In 2018, the comprehensive management

Y. Hao and T. Wan, *The River Chief System and An Ecological Initiative for Public Participation in China*, https://doi.org/10.1007/978-981-99-4921-2_1

of low-quality class V water was achieved, reaching the goal of eliminating the low-quality water that is three years ahead of schedule. In 2021, the water environment in Chongqing has improved significantly, 48 sections of black and odorous water in urban agglomerations have achieved "long-term regulation and long-term purification", the water quality of all sections of the national assessment in the mainstream of the Yangtze River has reached class II level, and the proportion of excellent water quality in the 42 sections included in the national assessment in the city has reached 100% for the first time, an increase of 11.9% over the period before the introduction of the river chief system in 2016 (She et al., 2019).

The long-term mechanisms of water protection have gradually established. Since the 18th National Congress of the Communist Party of China, China has revised and promulgated water management related regulations and laws, for instance the "Law of the People's Republic of China on the Prevention and Control of Water Pollution", the "Opinions of the CPC Central Committee and the State Council on Accelerating the Construction of Ecological Civilization", the "Action Plan for the Prevention and Control of Water Pollution", the "Regulations on Groundwater Management", and the "Opinions of the State Council on the Implementation of the Strictest Water Resources Management System". All provinces, autonomous regions and municipalities directly under the central government have established Offices of River Chiefs, introduced and improved working systems covering river and lake chiefs' task performance, supervision and inspection, evaluation and accountability, positive incentives. The RCS introduced and improved working mechanisms such as joint prevention and control, coordination and linkage, and social co-management of rivers and lakes (Wang & Chen, 2020). The ecological condition of rivers and lakes has changed historically, and a systematic water governance approach under the leadership of Chinese Communist Party and government has gradually taken shape.

An atmosphere of national participation in water management has been fostered. In the past five years, the organizational structure and responsibility system of China's River Chief System have been effectively introduced, not only improving the organizational structure of the five-level river chiefs to control water, with 300,000 river and lake chiefs at provincial, municipal, county and township levels and more than 900.000 village-level river and lake chiefs (including river patrollers and river guards), the river and lake chiefs at all levels assume the main responsibility for water control, coordinate local water management, achieve full coverage of river and lake management and protection responsibilities, and promote continuous improvement of river and lake management results. At the same time, the remarkable results of river chiefs in water management have aroused the enthusiasm of the whole society to participate in the management of rivers and lakes, and various models of citizen participation have emerged, such as the "foreign chief", the "enterprise river chief", the "village river chief" and the "women's river chief" have emerged, according to incomplete statistics, there are currently more than 7 million civilian river chiefs and volunteers for the protection of rivers and lakes in various places. The "Love Hundred Lakes" Volunteer Association of Wuhan City in Hubei Province not only led the lead in recruiting civil river and lake chiefs for Wuhan City, but also regularly carried out large-scale publicity activities for the common good, attracting more than 10,000 volunteers to participate in river and lake protection. They regularly cleaned garbage in rivers and lakes (Wan, 2021).

The comprehensive implementation of the River Chief System China's water governance has achieved remarkable results. Through the institutional design of the River Chief System, local governors act as river chiefs and bear responsibility for water management, which has promoted the reconstruction of the system of responsibility for water management. How is the responsibility of water management shaped by the River Chief System? From the responsibility perspective, the book provides an alternative agenda to deepen the understanding of the River Chief institutions System as a distinctive approach to solve water pollution and associated governance dilemmas.

2 Water Governance in China from a Historical Perspective

In China's centralized water control system, there have always been rich grassroots water control practices led by the "rice paddy governance model" (Hao, 2018), such as the primary alliance of adjacent land, the secondary association of the defense pond organization, and the advanced joint water control of the levee organization. While these voluntary mechanisms of joint water management played an important role on tributaries or in parts of the water system of rivers and lakes, they often failed during floods in large rivers or entire water bodies. Therefore, effective irrigation and flood control of large rivers and lakes must rely on control by central institutions (Fairbank, 1983). The Chinese bureaucracy is mainly there to meet the needs of the irrigation system. All governments in Asia cannot avoid performing an economic function, namely the function of carrying out public works. This artificial method of improving soil fertility is handled by the central government, which neglects irrigation and drainage, and the plant is immediately abandoned. Similarly, the first condition for the survival of China's agricultural civilization is artificial irrigation or river dredging. The realization of this condition is highly dependent on the implementation of the central government and its public duties to control water (Needham, 1974).

The central government's nationwide centralized water control has been particularly effective in the construction of major water conservation projects. For example, the construction of the Beijing-Hangzhou Grand Canal and the taming of the Yellow River depended on the state's implementation of public water control responsibilities in China's hierarchical system of water governance structure (Wang, 2018). The construction and maintenance of large-scale water conservancy projects often consume a lot of labor, materials, and financial resources, and high taxes, burdensome labor, and abusive penalties often make people dissatisfied and even exacerbate social conflicts. For example, the Sui Emperor called for more than one million migrant workers to build the Beijing-Hangzhou Grand Canal, which led to the outbreak of peasant revolts in the late Sui Dynasty. To repair the canal, in the ninth year of Yongle (1411), Zhu Di called for 600,000 households in Shandong to serve on the canal, which was nearly 75% of the total number of households in Shandong Province. The historical practice of water regulation in China shows that the construction of large-scale water conservancy projects is not normal. In the middle and lower reaches of the Yellow River, the middle and lower reaches of the Yangtze River, and areas of frequent flooding in the southwestern region, regional water control by local officials is the norm. In traditional societies limited by the people's ability to pay taxes, successive governments could govern the country with only as few officials as possible. In the late Qing Dynasty, 20,000 officials governed a country of 400 million people. The population administered by a single county official increased from 50,000 at the end of the Han Dynasty to 300,000 at the end of the Qing Dynasty. Regardless of the local governor's geographical area and population size, the local governor was responsible for the proper administration of finances, civil affairs, justice, and even military affairs within his administrative area. Especially for local officials who face severe flood disasters in their administrative areas, performing the public duty of water regulation is their main task (Fairbank, 1983).

In Chinese history, there have been many "river chiefs" who have managed water well. Besides the legend of water control by Dayu, the first river chief, there were many historical examples of water being managed by local river chiefs, such as LI Bing, the commander of Shu County in the Qin State; the construction of Dujiangyan; the dredging of West Lake by SU Dongpo in Hangzhou during the Song Dynasty; the management of the Yellow River by YU Qian, the governor of Henan and Shanxi provinces in the Ming Dynasty; and the irrigation of the Yellow River by LIN Zexu, the governor of Jiangsu in the Qing Dynasty. It was apparent that the people responsible for water management are local administrators. The local governor used the human, material and financial resources in the administrative area to effectively solve the flooding problems. These local chiefs were considered as excellent river chiefs at that time. Even more surprisingly, LIU Guangfu of Zhi County in Zhuji, Shaoxing, Zhejiang, faced frequent local floods, droughts and floods in the Ming Dynasty and used the system of responsibility to creatively design and implement the water management system achieving desired results. In traditional societies with relatively backward water management technology, local officials with the hierarchical responsibility system used limited social resources to maximize the transformation of floods into water conservation, promoting local social stability and economic development.

Despite the fact that in China major water conservancy projects were performed by the state power, water governance in rural areas was self-governed. The way of life in rural areas is generally determined by natural, social and agricultural conditions. Households sharing a well in neighboring areas is termed as "water connected community" by commons that are streams, lakes, rivers, ponds and wells (Qian, 2004). In the northern villages of traditional Chinese society, not only was a group structure formed around one or more wells, but also a public village square was built around a well, which are the source of drinking water for people and livestock, so they are an important cause of population agglomeration (Cao, 2006). In humid areas (South China, Jiangnan and other areas), villagers gather near rivers, lakes, ponds, dams, ditches, canals and other water sources to secure water for rice cultivation. Houses are scattered in small groups on the cultivated and farmed slopes and sparsely located in some gardens and "farmland", creating not only the structure of scattered villages, but also a public production space revolving around lakes, bogs, weirs, ponds, dams and so on. These dense water systems such as lakes, bogs, weirs and ponds not only form the basis for village development and the basic conditions for agriculture, but also invisibly shape the cooperation network of traditional society. Thus, a section of river often flows through several villages, and people from several villages form circles of cooperation beyond the villages because of water resources.

After the establishment of the P.R. China, the traditional production relations were transformed with the development of land reform and collectivization movement, the state power deeply interfered in the villages and villagers' lives, and the captain became the actual person in charge of water control at the grassroots. At this time, the countryside became cellularized coexistence, protection from drought and flood became a common need of the masses, and the captain's value consensus on water regulation became another day, and political dominance became the central force for shaping the public will for water regulation at the grassroots, which promoted the gradual formation of a political community for water regulation and realized the fundamental change of the basic conditions of water regulation in the countryside.

Especially after the people's communes entered the cooperative movement, the central government completed the restructuring of the grassroots organizations of rural society and created a new power structure. As a result, villagers are completely confined to the administrative unit (people's commune) controlled by village cadres, but at the same time, there are a large number of elite groups in the village and a large number of villagers' groups supporting collective action, which objectively provides strong authority of cadres, reliable personnel support and valuable action consensus for the collective construction of water conservancy. The rural cadres at that time, as agents of the state, used the two main tools of class struggle and socialist ideology to implement the will of the central authorities to the letter, so that the will of the central authorities was effectively transmitted and enforced (Siu, 1989). The agricultural production behavior of villagers was no longer determined by individuals, however it was subject to the unified order and command of the organization. As a result, the construction of water conservancy was no longer a voluntary cooperative act between peasants in the grassroots society, which was integrated into the state and becomes a highly politicized collective action under the leadership of the state, and at the same time it is incorporated into the nationalized villages, and various administrative means such as organizational mobilization, education and guidance, regulation and punishment, administrative orders and resource extraction have also followed and are widely used in the construction of water conservancy.

In the context of cellularized rural coexistence, the strong intervention of the state has profoundly reshaped the grassroots social governance environment and effectively adjusted people's patterns of thought and action. The key to improving water management after the establishment of the P.R.C. lay in the organization of the system, from inter-provincial planning to ditches in villages. Under such systematic organization, water management in villages had changed from a scattered voluntary association in the traditional society period to a unified centralized organization. In this context, the importance of the production team leader in water control was emphasized, who became the actual organizer and important person in charge of water control at the grassroots level. This centralized and unified form of organization integrates national will, concentrates forces at the grassroots level, mobilizes people's participation, coordinates a variety of resources, and has the remarkable characteristic of combining forces to do great things, which is also an important reason why rural water conservancies initiated large-scale development during the collectivization period. The positive response of politically engaged peasants to the Party and the government, as well as the educational and warning effect of various political movements, are not only a strong support for peasants' participation in the construction of rural water conservancies, but also an objective condition for maintaining cellularized villages.

For thousands of years, agricultural production has been "dependent on the heaven". Therefore, the pursuit of agricultural production to fight with drought and flood securing crops had become one of the greatest hopes of hundreds of millions of farmers. After the founding of the People's Republic of China, especially in the period of collectivization, the traditional water control community was restructured under the omnipotence policy, and the goals, methods and means of water control changed radically, with team leaders playing an indispensable role as representatives of the state. After the reform and opening up, in the face of the serious situation where resources are becoming scarce, the environment is seriously polluted and the ecosystem is damaged, based on the internal requirements of living community coexistence, co-ownership, consensus and public will, the leading river chief was created, responding to the public will to seriously carry out public responsibility.

With the progress of land reform and means of production redistributed to landless or landless villagers, it was too "evenly distributed" resulting in the dilemma of one family receiving 1/4 donkey and 1/4 cart, and more than a dozen households sharing one plow. Independent water farmers created a demand for cooperative water control, and seasonal and perennial irrigation organizations emerged in many places, such as "irrigation teams", "well drilling teams", "canal patrol groups", and "irrigation groups". The central government promoted the idea of agricultural cooperation and peasant organization, and governments at all levels also encouraged and supported the cooperation in water regulation that emerged in various localities, and promoted the establishment and operation of various mutual aid groups that were recognized by villagers. In view of this, after the completion of the land reform, the Central Committee of the Communist Party of China put agricultural cooperation on the agenda and adopted the Resolution on Mutual Aid and Cooperation in Agricultural Production in September 1951, calling on the entire Party to regard mutual aid and cooperation in agricultural production as an important matter, which was the beginning of the nationwide agricultural cooperation movement (Perkins, 2013).

In November 1955, the 24th Session of the Standing Committee of the National People's Congress deliberated and adopted the draft Model Statute for Agricultural Production Cooperatives, which defined peasant production cooperatives as collective economic organizations of working peasants. This established that members' private main means of production, such as land, livestock and agricultural tools, transferred gradually into public ownership. After the primary cooperative, it passed through the stages of advanced cooperatives and leapt into the people's commune. The people's commune is not only a production management organization, but also an administrative management organization, with the most dominant feature of implementing the commune ownership system internally and having the right to allocate and use human, material and financial resources for production teams and members. The private ponds, wells and other water management facilities were transferred to collective ownership. Collectivization of water management brought certain economies of scale, reduced water costs, safeguard crops during drought and floods, and promoted more efficient agricultural production (Chan et al., 1984). The collectivization of agricultural production has closely linked the interests of farmers, and the demand for the development of water conservation is no longer just the demand of each individual household, but begins to develop into a consensus of collective action, creating an ideological condition for the promotion of the construction of water conservation facilities at the grassroots level. After the start of collectivization, water conservancy projects, which were originally rare, became "commonplace", and the originally water-scarce farmland wanted to be developed into high-yield and stable-yield farmland, and the mountainous areas where there was originally no land were reclaimed by the mountains, and even artificial celestial rivers were constructed in the steep mountains. To implement these water conservation projects, water communities face complex social problems at various levels. Protection from drought and floods in collective agricultural production, the rapid construction of water conservancies compatible with collectivization, and efficient water management have become problems common to all members, and urgent practical solutions are needed.

In the period of collectivization, the grassroots organizational structure of political and social integration created the office of "river chief" which not only gave the team leader the power to direct water control, but also established the specific duties of the captain and river guide. Building water conservancy with the courage to "change the world for the earth" and improving the operating conditions of agricultural production are the basic judgments of the CPC Central Committee to increase crop yields. In the early days of the founding of the P.R.C., when the country was in ruins and the financial situation was very difficult, relatively large financial investments were still made in the construction of water conservancy. The investment in agriculture, forestry and water conservancy was 130 million yuan in 1950, 260 million yuan in 1951 and 640 million yuan in 1953, with an average annual growth rate of more than 100 percent, and the cumulative investment in three years reached 1.03 billion-yuan, accounting for 13.14 percent of the total capital construction investment of 7.84 billion yuan (Perkins, 2013).

After the completion of the "three major transformations", the need for water conservancy facilities for agricultural cooperative production increased, raising the problem of constructing water conservancy facilities to a new level. In October 1957, the Third Plenary Session of the Eighth Central Committee of the Communist Party of China adopted the "Forty Articles of the Agricultural Development Programme", which ushered in a peak in the construction of water conservancy facilities for agricultural land. On the one hand, agricultural production cooperatives dug wells, constructed ponds, built dams, drilled dry wells, opened canals, built polders, constructed reservoirs, built ditches and terraces for water storage and drainage, and treated small rivers according to local conditions. The construction of large and medium-sized water conservancy projects and the treatment of large and medium-sized rivers, on the other hand, should be carried out in a planned manner to avoid ordinary floods and droughts. For example, in the 1950s, lake ponds, lake dams, river dams and large river channels were dug and dredged in Songjiang County in the Yangtze River Delta region, and electric irrigation stations were established in almost every township in the 1950s (Huang, 1985). Water protection is a special task that requires a lot of human, material and financial support. In 1957, the Central Committee promulgated the decision of the CPC Central Committee and the State Council on launching a large-scale campaign to build agricultural water supply and accumulate fertilizer this winter and next spring, which clarified the principle of water supply under the leadership of the government. This method of water supply fully aroused the enthusiasm of farmers and was not only effectively promoted throughout the country, but also continued continuously throughout the collectivization period. Relevant data show that in 1957, the effective irrigated area of arable land in China was 410 million mu, accounting for 24.4% of cultivated land, while in 1978, the effective irrigated area of arable land increased to 670 million mu, accounting for a net increase of 260 million mu, and the proportion of cultivated land increased to 45.2%, accounting for an increase of 20.8% (General Office of the Standing Committee of the National People's Congress, 2007).

Compared to the water conservancy projects constructed by grassroots communities in the traditional society period, the water conservancies completed by the state are completely different from the customary law in terms of daily use and maintenance. All completed water conservancy facilities are the responsibility of rural grassroots organizations such as communities, production brigades and production teams. There are clear rules about when they can be used, who makes the general arrangements, who is responsible for operation, and what specific responsibilities and duties exist. In the modern state, there is not only the vertical intervention of administrative power in rural society, but also the flexible use of national political discourse and respect for local logic, such as stimulating peasants' enthusiasm through bitter thoughts and sweetness. As agents of the state in the villages, these specific leaders strictly execute the will of the state and shape the relationship between the state and its members. Even members managed water resources following the state's rules and developed a new "collective sense of life".

3 Principal-Agent Relationship, Responsibility and State-Society Mutual Empowerment

From the perspective of the principle-agent theory, there are several interactive relationships between the society and the government, between the central government and the local government, and between the upper and lower levels of local governments in the grassroots water environment management process. The introduction of the River Chief System has strengthened the responsibility of local officials for water management and principal-agent relationship. The asymmetry of information requires a better understanding of the responsibilities of local officials for water management and increased monitoring of the fulfilment of the responsibilities of local officials for water management. In the context of the establishment of the River Chief System, local officials assume responsibility for water control entrusted to them by the people, which is the reproduction of the principal-agent relationship in the field of environmental protection and conservation. By the delegation of consent authority, the public expresses their demands for a better ecological environment in the form of public opinion, and transfer responsibility for water management to local governments. As the representative of the people, the government is the executor of sovereignty and should act in accordance with the demands of public will and respond to public will as much as possible (Mill, 2015). In exercising its power, the government must abide by the contract concluded between it and society, earnestly fulfill the corresponding rights and duties, and not forget its responsibility to society (Kassim & Menon, 2003). It is considered that the process of implementation of the public will by the government is the process of responsibility to the authorized subject as the public.

From the perspective of the government hierarchy, the public, the central government, local governments, and other plural subjects form a chain of principal and agent. This shapes multiple principal-agent relationships between the public and government agencies, between the central government and local governments, and between local governments and subordinate agencies. As the representative of the people as a whole, the central government not only bears the heavy responsibility for national ecological water management, but also has the task of supervising local governments in the active performance of their water management duties. As the representatives of local people, local governments not only bear the responsibility for ecological water management in their jurisdiction, but also actively fulfill the supervision requirements of higher levels of government, the central government and the public. In the context of the establishment of the River Chief System, the formation of the principal-agent relationship between the people and the government establishes the legitimacy of the local government's water management responsibility, and the active performance of water management duties by local officials is an affirmation of this principal-agent relationship and a strong implementation of the will of the principal.

Against the backdrop of the establishment of the River Chief System, the overlay of multiple principal-agent relationships has made information asymmetry increasingly severe, requiring increased oversight of water management by local chiefs.

In terms of water environment management, China applies the principle of block management, which leads to an extension of the management chain and an increase in governmental distance, thus increasing information asymmetry unnoticed. Increasing information asymmetry means that the central government is not sufficiently aware of the responsibilities of local officials in water management and that local governments can make more strategic decisions. The local river chiefs as street bureaucrats, who have discretionary power, can use their structural advantages to selectively report water management problems to consolidate their information power and the associated safety valve effect. From the perspective of the relationship between the central government and local governments, local governments can protect grassroots' interests of water management due to different interests and the influence of asymmetric information factors. Under the impression of seeking short-term political gains, some local governments may enter into collusion with highly polluting enterprises, and the supervision of some of the agencies responsible for pollution control is not strong enough, resulting in tokenistic implementation of water environment policy, which not only makes it difficult to form a benign pattern for long-term water control, but even worsens the governance ecology of the local water environment.

In addition, from the perspective of interactive relationship between the public and the government, information asymmetry in the process of water control may make it difficult for the public to monitor the government's water environmental policies. For example, in some areas, there are problems such as inadequate disclosure of information, insufficient information channels and poor interpretation of information content, which may dampen the public's enthusiasm to participate in water management to a certain extent and affect the public's sense of gain. Therefore, in the context of the establishment of the river chief system, due to the problem of information asymmetry caused by the overlapping of multiple principal-agent relationships, it is objectively necessary to strengthen the effective supervision of the fulfilment of local officials' responsibilities for water management, in order to avoid and prevent the moral hazard and adverse selection caused by information asymmetry. With the scientific establishment and precise determination of the responsibilities of local officials in water management, effective supervision of the responsibilities of local officials for water management has become a practical requirement to solve the problem of water management and promote high-quality development of the ecological environment.

Accountability is both an important variable in public administration and a key factor in the efficient functioning of administrative organizations. Therefore, clear rights and duties, as well as clear responsibilities, have also become one of the basic principles of organizational design within the bureaucratic system. The question of how to scientifically design and effectively allocate the vertical and horizontal responsibilities of government has always been an important issue that cannot be ignored in modern local governance. It is pointed out the dilemma of floating responsibility in the modern administrative system (Bauman, 1994), illustrating that an actor of the organization transfers his responsibility to his superiors and he himself falls into the agent state of executing orders, which causes the organization to float and suspend responsibility when transferring responsibility and evading responsibility.

In this sense, eliminating the floating of responsibility and realizing the reconstruction of responsibility through accurate responsibility definition is an important task of modern administrative management.

4 Water Pollution, Whose Responsibility, and Innovating Water Governance Institutions

Water is the source of life, the cradle of civilization, and the foundation of development. With the rapid economic and social development, combat and prevention of rivers and lakes pollution has become an acute problem in China, which have brought formidable challenges to ensuring water security. In front of the increasingly server water pollution and complex water governance context, it is difficult for the "nine dragons controlling water" mode to avert the "tragedy of the commons". The "nine dragons controlling water" mode demonstrates Chinese *Tiara-Kuai* governance structure. *Tiao* is the vertical authority from top-down and *Kuai* refers the horizontal authority of territorial governmental relations (Mertha, 2005). The issue of responsibility sharing among local governors has long been unclear, concerning governmental functions and water pollution prevention responsibilities. The overlapping of responsibilities between central and local governments often lead to *Tiao's* 'no responsibility' and *Kuai's* 'no burden' in water governance. The local governors do not share both responsibilities and burdens for the central government. Apparently, the *Tiao* and *Kuai* water governance modes are in dilemmas.

The *Tiao* governance mode sees the absence of responsibility of local officials in water governance, due to responsibility isomorphism. This is because of the influence of historical and cultural traditions, state's unitary structure, and planned economy. China's high degree of unity in the functions, responsibilities, and institutional settings of different levels of government in the *Tiao* vertical line has shown the characteristics of 'top-down', 'upper and lower counterparts', and 'left–right alignment', which is understood as 'isomorphism of responsibilities' (Zhou, 2009). In water pollution control, the central government and the local government take part in it as the main managerial bodies. However, the central government, which has overall and absolute control power over local governments, a pole to the end in the figurative sense, not only has difficulty in establishing an effective hierarchical governance system, but also creates an overlap of macro-regulation function and micro-management function between the central and local levels of governments. On the one hand, the lack of a clear responsibility clarification between the central government and local governments at all levels in water pollution control resulted in unclear boundaries of responsibilities among governments; one the other hand, it causes a dilemma that in water pollution local governmental heads do not need to take responsibilities when they are not due diligence.

The vertical intergovernmental relationship of 'responsibilities isomorphism' has plunged local governmental heads into the dilemma of 'local totalism' and 'local inability'. Although the work and tasks faced by local officials at all levels are the same as those of the higher governments in the local totalism sense, they are constrained by the limited power of accessing resources showing local inability. Local governments oscillate between 'local totalism' and 'local inability' (Liu et al., 2016). Especially under the batons of performance appraisal, local governors pay more attention on how to further enhance GDP and achieve better political performance, rather than putting much effort into dealing with water pollution. When they undertake tasks of water pollution prevention, it is difficult for them to do due diligence as their primary focus is guided by performance appraisal.

Furthermore, several departments involved in Water pollution prevention and control in China involve several departments, for instance Water Conservancy Department, Environmental Protection Bureau, Bureau of Housing and Urban–Rural Development, Development and Reform Bureau, Department of Natural Resources, Health Bureau, Bureau of Agriculture and Rural Affairs, Economy and Trade Bureau, etc. They are under the leadership of and accountable to both local governors and various vertical departments. When the local water pollution control is not effective, though the local governors are accountable for this, they would shirk part of their responsibility towards the departments in the vertical line. Therefore, under such a governance system, local governors are not motivated to take responsibilities in water pollution control.

The *Kuai* governance mode is also challenged by the dilemma that how to hold local chief officials accountable in water pollution control (Zhu & Chen, 2016). The failure of local chief officials' taking responsibilities for water control due to a vital fact that the boundaries of rivers and administrative districts are not overlapping, though it is clearly stipulated in the law on "Environment Protection Law of the People's Republic of China" enacted in 1989, stating that local government at all levels shall be responsible for the environmental quality of their jurisdictional areas and take actions to improve the environmental quality. Not only does Articles 4 and 5 of the revised "Law of the People's Republic of China on Water Pollution Prevention and Control" promulgated in February 2008 stipulate that "local government at or above the county level shall take countermeasures and measures to prevent and control water pollution, and shall be responsible for the water environment quality of their respective administrative regions", but also point out that "the State implements the target responsibility system, assessment and evaluation systems for water environment protection, taking the completion of water environmental protection objectives as the content of assessment and evaluation for local governments and chief officials" (Ministry of Ecology & Environment of the People's Republic of China, 2008).Within the water pollution related legal framework, the local governors should bear the responsibility of water pollution control in their jurisdiction.

When a river flows through different administrative areas, there are a series of conflicts and conflicts among local governments in water resources development, water utilization and water pollution prevention and control, concerning arguments over upstream or downstream, left banks or right banks, and main stream or tributaries (Fan, 2015). Attention given on the local economy development, local governors whose jurisdictional areas are located in the upstream areas not only intercept a large amount of water, leading to a sharp decrease in the flow of the river, seriously affecting the demand for water in the lower reaches, but also overexploit and utilize the rivers within their jurisdiction. These cause ecological damage to rivers and restrict water utilization and water resources development in the downstream areas. In addition, because of local protectionism the short-term behaviors of some local governors lead to ‘prisoner’s dilemma’ of non-cooperative games between local governments (Yu & Yang, 2021). Besides, China’s *Tiao* and *Kuai* compartmentalized governance system not only makes difficult to solve problems of compensation and pollution liability attribution for trans-regional sewage treatment, but also provokes further conflicts between upstream and downstream, left and right shores, main streams and tributaries in the coordination of local interests, upstream and downstream ecological protection and ecological compensation mechanisms (Li & Hu, 2017).

Despite the central government has made special legislation requiring local governments to coordinate the control of water pollution, local water governance shows the characteristics of fragmentation in terms of resources management, local governors’ powers usage and policy formulation and implementation at local level. In addition, there are various obstacles and ineffective information circulation among local governments, weaknesses in ecological compensation mechanisms, management systems and related technical operations. It is difficult to hold local governors accountable for trans-regional water pollution governance, highlighting the urgency of reforming water governance mode and mechanism.

Moreover, the terrain feature in China is high in the west and low in the east, showing a three-step ladder terrain feature with crisscrossing mountains, complicated landforms, and diverse climates, and reticulated rivers. Many rivers flow across county boundaries, city boundaries and provincial boundaries. According to the survey results of the National Bureau of Statistics, the Yangtze River, the largest river in China, accounts for one-fifth of China’s land area and flows through 19 provincial-level administrative regions (National Bureau of Statistics of China, 2021). This poses the challenge of defining the property rights of rivers that flow across regions, which are critical to understand water resources governance (Agrawal, 2001; Schroeder & Castillo, 2013; Wang et al., 2017). The biophysical boundary of rivers is easier to define via technical means than demarcating responsibilities for water pollution prevention and control.

Water pollution prevention and reduction has become an important task in China when economic development provides impetus for advanced water demand emerging from stakeholders. Nevertheless water environment and water ecological system are downplayed (Wang et al., 2018). Local governors are supposed to play the key role in averting the “tragedy of the commons” and coordinate the “nine dragons controlling water”, which are attributed to transboundary nature and ambiguous property rights of water resources. Arguably, the biophysical conditions and complex social attributes of water resources catalyze innovated river governance institutions, River Chief System, denoting that river chiefs are local governors, including governmental heads and Communist Party Leaders at all local levels, who are the subjects of responsibility for water pollution prevention and control. In other words, River Chief System is considered as a responsibility contracting system.

References

Agrawal, A. (2001). Common property institutions and sustainable governance of resources. *World Development, 29*(10), 1649–1672. https://doi.org/10.1016/s0305-750x(01)00063-8

Bauman, Z. (1994). *Alone again: Ethics after certainty*. Demos.

Cao, J. (2006). *China along the Yellow River: Reflections on rural society*. Routledge.

Chan, A., Madsen, R., & Unger, J. (1984). *Chen village: The recent history of a peasant community in Mao's China*. University of California Press.

Fairbank, J. K. (1983). *The United States and China: Revised and enlarged* (Vol. 10). Harvard University Press.

Fan, J. (2015). The criminal system of environmental pollution on the basis of water pollution (in Chinese). *Journal of Henan Normal University (philosophy and Social Sciences Edition), 42*(3), 39–79.

General Office of the Standing Committee of the National People's Congress. (2007). *Compilation of documents of the Fifth Session of the Tenth National People's Congress of the People's Republic of China*. People Press.

Hao, Y. (2018). "Rice Field Governance Model": Grassroots water autonomy in China's water control system. *Political Research* (in Chinese) (4), 48–57.

Huang, P. (1985). *The peasant economy and social change in North China*. Stanford University Press.

Kassim, H., & Menon, A. J. J. o. E. P. P. (2003). The principal-agent approach and the study of the European Union: Promise unfulfilled?, *10*(1), 121–139.

Li, Y., & Hu, Y. (2017). Analysis on the path of holistic governance of river basin ecological environment: Based on river chief system reform (in Chinese). *Research on Socialism with Chinese Characteristics* (4), 73–77.

Liu, F., He, T., & Zhou, Y. (2016). Analysis of the rule of law of the "River Chief System" in the context of governance modernization (in Chinese). *Zhejiang Academic Journal* (6), 120–123.

Mertha, A. C. (2005). China's "soft" centralization: Shifting Tiao/Kuai authority relations. *The China Quarterly, 184*, 791–810. https://doi.org/10.1017/S0305741005000500

Mill, J. (2015). *An essay on government*. Cambridge University Press.

Ministry of Ecology and Environment of the People's Republic of China. (2008). *Gazette of the Standing Committee of the National People's Congress of the People's Republic of China*. Beijing General Office of the Standing Committee of the National People's Congress.

National Bureau of Statistics of China. (2021). China statistical yearbook. China Statistics Press. http://www.stats.gov.cn/tjsj/ndsj/2021/indexch.htm

Needham, J. (1974). *Science and civilisation in China* (Vol. 5). Cambridge University Press.

Perkins, D. H. (2013). *Agricultural development in China, 1368–1968*. AldineTransaction.

Qian, H. (2004). Xianghu lake water conservancy: The institutional foundation of the commons. *Zhejiang Social Sciences* (in Chinese) (6), 163–169.

Schroeder, N. M., & Castillo, A. (2013). Collective action in the management of a tropical dry forest ecosystem: Effects of Mexico's property rights regime. *Environmental Management, 51*(4), 850–861. https://doi.org/10.1007/s00267-012-9980-9

She, Y., Liu, Y., Jiang, L., & Yuan, H. (2019). Is China's River Chief Policy effective? Evidence from a quasi-natural experiment in the Yangtze River Economic Belt, China. *Journal of Cleaner Production, 220*, 919–930. https://doi.org/10.1016/j.jclepro.2019.02.031

Siu, H. F. (1989). *Agents and victims in South China: Accomplices in rural revolution*. Yale University Press.

Tian, J., Han, Y., Shen, J., & Zhu, Y. (2022). Leveraging sustainable development of agriculture with sustainable water management: The empirical investigation of "Five Water Cohabitation" of Zhejiang Province in China. *Environmental Monitoring and Assessment, 194*(2), 124. https://doi.org/10.1007/s10661-022-09771-6

Wan, L. (2021, January 18). Relay of love! Three Generations of the Wuhan Family Guarded the Lakes. *Yangtze River Daily*. http://swj.wuhan.gov.cn/tzdt/hhzz/202101/t20210118_1599788.html

Wang, Y. (2018). Hierarchical structure of water governance. In Y. Wang (Ed.), *Assessing Water Rights in China* (pp. 31–62). Springer Singapore.

Wang, Y., & Chen, X. (2020). River chief system as a collaborative water governance approach in China. *International Journal of Water Resources Development, 36*(4), 610–630. https://doi.org/10.1080/07900627.2019.1680351

Wang, Y., Wan, T., & Biswas, A. K. (2017, October). Structuring water rights in China: A hierarchical framework. *International Journal of Water Resources Development*. https://doi.org/10.1080/07900627.2017.1378627

Wang, Y., Wan, T., & Tortajada, C. (2018). *Water Demand Framework and Water Development: THe Case of China, 10*(12), 1860.

Yu, M., & Yang, X. (2021). How to "Take the neighbor as the enemy" towards "Sharing the same bed": A study on cross-administrative cooperative governance based on the Yellow River Basin (in Chinese). *Public Governance Research, 33*(6), 5–13.

Zhou, Z. (2009). *A study of the "Tiao-Kuai" relation of the contemporary Chinese Government* (in Chinese). Tianjin People's Publishing House.

Zhu, H., & Chen, C. (2016). The moral foundation of environmental ethics (in Chinese). *Journal of Hubei University (philosophy and Social Sciences Edition), 43*(5), 39–79.

Chapter 2
Top-Down Mechanisms: Governmental River Chiefs

1 Politics of Responsibility and Responsiveness

The politics of responsibility is the basic feature of modern democratic politics, and it is also the internal logic that must be followed to promote the system of River Chief System. To realize the politics of responsibility, it is necessary to observe the coherence of rights and duties in the allocation of power, to strengthen the supervision of responsibility in the exercise of duties, and to evaluate and hold accountable the effects of the exercise of power (Hao, 2017b). Regarding the River Chief System, according to the requirements of the responsibility policy, it is necessary to effectively clarify the power allocation of river chiefs, standardize the fulfillment of river chiefs' responsibilities, and improve the effectiveness of river chiefs' water management.

Coherence of power and responsibility is an important principle of accountability policy in the allocation of power, reflecting the requirement that the power of the office must match the responsibility. In theory a government can win the recognition and support of the people only if it fulfills its duties, is faithful to its responsibilities, and has the courage to take responsibility (Linde & Peters, 2020). However, in the course of grassroots governance, unequal rights and responsibilities have become normal practice. The absence of law, the expansion of power, and the pressure system have become important cognitive tools to understand and grasp this phenomenon. The theory of the absence of law states that the imbalance of rights and obligations is due to the absence of appropriate institutional arrangements such as legal norms in the allocation of rights and obligations. The concept of power expansion emphasizes that power has the property of expanding, and in the case of profit motive and lack of constraints, it inevitably leads to an imbalance of power and responsibility. The pressure system theory indicates that the transmission of pressure at the lower levels of the pressure system has shaped the power and responsibility pattern of the government at the grassroots level, resulting in an imbalance of power and responsibility. Therefore, how to resolve the imbalance of power and responsibility in the work of grassroots governments has become an important issue to promote the transformation of government functions and build a responsible government.

Y. Hao and T. Wan, *The River Chief System and An Ecological Initiative for Public Participation in China*, https://doi.org/10.1007/978-981-99-4921-2_2

Concerning River Chief System, the process of water management by local chiefs should also follow the principle of consistency of authority and responsibility. Since the River Chief System gives local chiefs full authority to manage water, the responsibilities of river chiefs must be divided scientifically. For example, from the perspective of building the responsibility system, political responsibility, administrative responsibility, legal responsibility and technical responsibility are the main meanings, and local officials should not only adhere to the joint responsibility of the party and the government and bear the political responsibility for ecological water management, but also exercise the administrative responsibility and be the first responsible person for water conservancy in the region. It is not only to bear legal responsibility for the protection of the water environment, but also to exercise professional responsibility and promote the professionalization, standardization and modernization of the management and protection of rivers and lakes.

In terms of the way of constructing responsibility, the basic way of constructing responsible water management is the basic way that includes both the main responsibility of the party and the government, the hierarchical responsibility supported by the administrative level, and the segmented responsibility limited by the same river basin. From the perspective of defining responsibility, the list of responsibilities of river chiefs and the division of responsibilities in legal form explored in practice help to clarify the water management responsibilities of local chiefs from an institutional perspective and clearly delineate the responsibility relationship between local chiefs and water-related departments. For example, the newly revised "Water Law of the People's Republic of China" stipulates that China's water resources management should implement the method of combining river basin management and regional administrative management, thus strengthening the responsibility of local water management in law and providing a legal basis for river chiefs to actively carry out their duties. In addition to formulating a unified list of local powers and responsibilities, Beijing has also formulated a list of specific responsibilities for special issues such as environmental protection to provide a basis and reference for the scientific definition of local officials' responsibilities for water management.

From the perspective of the work process of public administration, strengthening the supervision of responsibility is the real meaning of the political theory of responsibility on the idea of power control. Political science generally recognizes the following three mechanisms to limit power: first, it limits power through morality and focuses on preventing abuse of power by strengthening the moral self-discipline of those who exercise power. Second, it seeks to limit power through power control to prevent abuses of power. Third, to limit power through rights, focusing on protecting the rights of citizens and preventing abuses of power. Therefore, strengthening the supervision of the exercise of power and regulating the exercise of power is always an important issue to ensure that power holders perform their duties in a consistent manner.

In the context of establishing the system of river managers, it is necessary to unify the performance of water management duties by local managers, from the perspective of responsibility supervision. In accordance with the idea of limiting power with power, the river chief system realizes the hierarchical control of the river chief's responsibility for water management in the river basin through the implementation of local responsibility, river basin classification and segmentation, vertical evaluation and accountability, etc., and strengthens the process supervision of the performance of river chiefs' exercise of responsibility for water management. At the same time, in line with the idea of limiting power through morality, the implementation of the river chief system focuses on strengthening the responsibility ethics of local officials in water management, which is integrated into the framework of building political ethics, and constantly strengthens the sense of responsibility and action of local officials in water management. In addition, it is necessary to further improve the laws and regulations related to the river chief system by relying on the governance proposition of limiting power through rights, which not only provides a direct legal basis for the effective operation of the River Chief System and water management by the river basin manager in accordance with the law, but also encourages citizens to participate in the supervision of river chief by improving the basic rights of the public in local water management. Under the background of the establishment of the river head system, the endogenous theoretical appeal of the responsibility policy for the supervision of river chiefs' responsibility will lead local governments to explore and strengthen the supervision of river chiefs' responsibility for water management, and promote the formation of a three-dimensional supervision network of moral constraints, power constraints and legal regulations.

The politics of responsibility not only emphasizes coherence of power in terms of source and process control, but also pays more attention to monitoring and evaluation. Moreover, the political theory of accountability clarifies the operational basis for conducting assessment and accountability, focusing on governance outcomes and improving organizational performance (S. Wang & Yao, 2007). In the River Chief System, local officials need to pay more attention to the actual results of water control practice. In short, the actual water management effect of local officials is not only an important part of evaluating the performance of local officials' ecological protection functions, but also an important basis for holding local ecological and environmental protection functions accountable for their unfavorable performance. Evaluation and accountability are the baton and the wind vane that are of great importance in guiding local officials' performance in water governance. However, the evaluation and accountability system of China's ecological civilization still has some shortcomings, such as the biased evaluation indicators, the virtual contents of evaluation, the insufficient participation of the public in evaluation and the inadequate utilization of evaluation results.

In the implementation of the River Chief System, in order to strengthen the active exercise of local chiefs' responsibility for water management, it is still necessary to pay attention to the construction of the system, the organization and implementation of assessment and accountability, and the application of the results, so as to truly give full play to the institutional power of assessment and accountability. First, how to strengthen the structure of the evaluation system and the establishment of indicators to better reflect and truly reflect the reality of grassroots water control and recognize whether the outstanding problems in grassroots water control have been fully addressed. It is crucial to achieve a balance between hard and soft indicators.

The second question is how to better evaluate the organization and implementation of accountability, so that more people can participate in the evaluation of the impact of water management through various forms of system design and mechanism innovation, and truly assess and understand whether it meets the public's demand in the practice of water management. The third question is whether the application of the results of evaluation and accountability through the pressure transmission mechanism created by evaluation and accountability can standardize the performance of local officials in water management and even transform the political ecology of local water management and create an effective water management pattern of party committee leadership, joint management, departmental coordination and public participation. Therefore, with the help of the pressure transmission mechanism, result-oriented evaluation and accountability gradually promote local governments to regard water management as an important ecological function, implement it as a central task, and serve as an important responsibility of local officials, which also provides a theoretical basis for the scientific definition of local officials' responsibilities for water management.

Ecological water management reveals an expression of public opinion. The government's responsiveness indicates the government's capability of good governance. This inherently requires the government to attach importance to the expression of the public's demands, regards the response to the people's public opinion as its own responsibility, and continuously improves its ability to respond positively. In the context of establishing the river head system, the government must pay attention to the people's demands for ecological water management in response to public opinion, regard the government's water management as an important embodiment of improving its own response capability, and strengthen the government's response capability through the senior official in charge of water management.

Since ancient times, China has been facing the basic national conditions of extremely unequal distribution of water resources in time and space, frequent floods and droughts, which forces the government to carry out public responsibilities to control floods. Therefore, Chinese civilization and water management are closely related, as it is argued that the origin of Chinese civilization is related to water management, the characteristics of Chinese civilization are those of water management civilization (Perdue, 2013). With the establishment of the People's Republic of China and the continuous improvement of state governance capacity, floods such as those of the Yangtze River, Yellow River, Pearl River and Huai River could be controlled quickly and effectively. However, rapid economic and social development in various places has also led to water pollution, damage to the water environment and degradation of water ecology due to lack of environmental awareness, seriously affecting the safety of people's lives and property. In many places, previous problems of water ecology have not yet been solved, and new problems have gradually appeared, even extremely harmful.

Water resources are commons characterized as high competitiveness and low exclusiveness (Ostrom, 1990). Even with tighter ecological water management, the nature of public affairs itself still leads to the tragedy of the commons, the prisoners' dilemma, and the dilemma of collective action. Self-governance is suggested by Ostrom effectively circumvents the dilemma of collective action, it is difficult to be effective in the face of governing large-scale natural water resource. From the internal logic of water control, there is a significant difference between river chiefs managing water under the leadership of local officials and the self-governed water management. The river chief represents the will of the public and the self-organization represents the will of the people. The will of public represents the common interests and demands of the people as a whole, is a concentrated embodiment of public interests, and therefore can serve as a fundamental basis for governments at all levels to formulate and improve policies (W. Li et al., 2012). At the 62nd National Congress of the Communist Party of China, the Party Central Committee headed by Comrade Xi Jinping attached great importance to the construction of ecological civilization and regarded it as an important part of the overall concept of "five-in-one" and the coordinated promotion of "four comprehensive strategies" (He, 2019). In the new era, people's demand for a growing and better ecological environment, such as the APEC blue, people have strong demands for smog control and have maintained a high level of attention to the three major battles launched by the Party Central Committee. Another example is people's ever-increasing interest in healthy drinking water, which will increase the sales of water purifiers from 5.75 million units in 2012 to 18 million units in 2020. With the growing public demand for ecological environmental protection, the environmental awareness of Chinese citizens has gradually awakened, and all kinds of non-governmental environmental protection organizations have also initiated rapid development and standardization improvement. According to the relevant data from the Ministry of Civil Affairs, by the end of 2016, there were 6444 ecological and environmental organizations in China, playing an important role in environmental policy participation. Therefore, the growing demands of the people for environmental protection and the increasingly complex and deteriorating ecological conditions of water bodies urgently require the government to fully and effectively incorporate the interests and general will of the broad masses of the people and respond to them in a timely manner, and actively implement the public function of water control.

Publicity is the fundamental attribute and principal property of government that establishes and constrains government policy-making and government action, i.e. the government should go beyond narrow self-interest, formulate and implement public policies in accordance with the common interests of society and the will of the people, from the fundamental point of view of ensuring the interests of citizens. From the theoretical point of view, the public will as the expression of the concentrated will and interests of the whole people should become the fundamental basis for the government in formulating policies, an important factor influencing the basis of legitimacy, and optimize the value orientation of policies (Rawls, 2020). The Chinese government's implementation of the public function of water management is the epitome of fulfilling the public's demand for ecological water management.

From the perspective of the main body of ecological water management, Chinese river water management is still a national water management. Ostrom believes that state-imposed water management plans are often oversimplified or idealized

(AGRAWAL & OSTROM, 2001), such as how central institutions should be organized, what powers they should have, how those powers should be limited, and how information should be obtained, how agents should be selected, how their work should be encouraged, monitored, or sanctioned, etc., which are not clear enough to implement. At the same time, the state carried out a nationalization that transformed previously restricted public property into an open resource. The nationalization of forests that were previously collectively owned had disastrous effects. In government-led river water management, not only have the functions of governments at all levels been clarified, but the main responsibilities of the Party's and government's key leaders have also been clarified, and information-sharing platforms and incentive and control mechanisms have been established to meet the new requirements of ecological water management. When local governments perform the public functions of water control, the remarkable results of the policy have won the endorsement and support of the "public forces" and form a joint force for water control.

In the new era, the central government has strengthened the responsibility of local governments for ecological water management in an unprecedented way, which is accompanied by the pressure of monitoring, evaluation and accountability brought by the normalization of environmental protection inspections. Having felt the pressure from their superiors and the demands of the people, local governments are actively adjusting their ecological and environmental protection behaviors, making water protection an important part of their daily work and even one of the central responsibilities of local governments. From the perspective of government response, this change means that local governments recognize the importance and urgency of ecological water conservation, actively consider ecological water management as their own responsibility and task, and actively respond to public opinion on ecological water conservation. When the government actively responds to the people's interests and effectively meets them, the people's happiness will be enhanced, and the government performance will be improved and strengthened. From the perspective of the impact of the government's response, the government's positive response can create the necessary information conditions and mechanism channels for citizen participation, effectively improve the sense of effectiveness of citizen participation, and then stimulate the public's growing enthusiasm for participation in water management, fundamentally change the problem of insufficient citizen participation in water management in China, and promote the formation of a benign pattern of coordinated water management. Therefore, in the context of the establishment of the river head system, the government consciously performs the function of water control, actively perceives and responds to public opinion on water control, which is a concrete embodiment of the government's responsiveness.

Local party and government leaders are administrative heads with administrative powers and control over government resources, and their capacity for resource coordination and political mobilization is unprecedented at the same level. Under the background of the introduction of the river head system, with the help of positioning the responsibility of local chiefs for water management, the main responsibility of river heads for water management is effectively clarified, and the vertical governance responsibility system of local chiefs for water management is built by transferring the pressure down layer by layer. From the point of view of horizontal power exercise, with the help of administrative authority and coordination and integration ability of local officials, the enthusiasm of various departments for coordinated water management can be effectively mobilized. In practice, some localities have experimented with the establishment of river management committees whose members are not only river managers who are officers of the Party and the government, but also officers of various functional departments, in order to clarify the division of labor among departments, strengthen interdepartmental coordination, share information resources and play out the advantages of each department, while coordinating and mobilizing local water control forces as much as possible.

It should be pointed out that local chiefs have the advantages of administrative powers and administrative resources, so they have the basic conditions to take responsibility for water management, but whether they can take full advantage of local chiefs' responsibility for water management is closely related to factors such as local officials' response consciousness, response capability and corresponding response mechanisms. From a response awareness perspective, local officials' response awareness directly affects their perception of the public's demands for water control. Response consciousness is often closely related to cadres' conception of policy performance, and an incorrect conception of policy performance leads to a lack of response consciousness among local officials. Therefore, in order to strengthen the response consciousness of municipal officials, a people-oriented concept of political performance must first be cultivated and established. From the perspective of responsiveness, responsiveness is directly related to the practical effect of local governments in responding to citizens' demands. Government responsiveness is a key element of good governance and an important measure of how well a country or society is governed. In order to strengthen the responsiveness of local officials, it is necessary to first cultivate party spirit, conduct training for enterprises and strengthen accountability supervision. From the perspective of the response mechanism, a sound response mechanism is an important guarantee that the government can actively respond to the public's demand for water control and conscientiously fulfill its functions and responsibilities. In the context of establishing the river chief system, from the government's point of view, the river chief who acts as the local leader should lead the water management function, and it is also necessary to strengthen the local leader's response awareness, response ability and response mechanism, and continuously promote the river chief to actively carry out the functions and responsibilities of water control.

2 Responsibility-Driven Leadership

The chief river head of each province, autonomous region or municipality directly under the central government shall be the first person responsible for the management and protection of the rivers and lakes in its administrative area and shall have overall responsibility for the management and protection of the rivers and lakes and shall bear the primary responsibility; River chiefs at other levels are the persons directly responsible for the management and protection of the corresponding rivers and lakes, and are responsible for the graduated and segmented management and protection of the corresponding rivers and lakes. In this way, a vertical system of contractual responsibility has emerged with subject responsibility as the core and hierarchical responsibility and segmented responsibility as the two main forms (Hao, 2017b).

Xi Jinping chaired the 14th meeting of the Central Leading Group for the Comprehensive Deepening of Reform, pointing out that the main inspectors implement the decisions and commitment of the Party Central Committee, solve major environmental problems and exercise the main responsibility for environmental protection. It is necessary to strengthen the requirements of 'equal responsibility of the Party and the government' and 'dual responsibility of an office' for environmental protection, and to investigate the responsibility of relevant units and individuals in places where problems arise. Local Party and government leaders bear the main responsibility for building ecological civilization in their jurisdictions and are the "first in charge" of water environment management in their administrative regions, and should improve the efficiency and level of water environment management.

Through definition, distribution, operation, evaluation, etc., the system of river managers promotes party and government representatives to take the main responsibility for water management. First of all, the main responsibility of party and government representatives for water management should be precisely divided. Traditional river (lake) management issues and responsibilities are intertwined with other responsibilities, difficult to separate, and rarely become the preferred choice of the government's responsibility system, such as sacrificing the environment to promote economic development in many places. A study discusses the experience of successfully implementing the river stewardship system in the city of S. The first task in establishing the river stewardship system is to clearly define and divide the responsibility for water management, and to clarify the responsibility for water management and its position in the jurisdictional system (Han & Li, 2021). The City of S defines water management responsibilities primarily along two dimensions: First, spatially, in accordance with the principle of territorial management, all issues of management within a certain area of rivers (lakes) and their certain areas are considered to be responsibilities for water management, such as requiring that there should be no buildings within 6 m of the city river (lake) that are not for water management; second, in terms of time, defining the responsibility for the whole cycle, defining the secondary problems caused by poor water management as responsibilities for water management, and studying the responsibilities of the relevant responsible parties, such as the reduction of crop production and extinction caused by the pollution of rivers (lakes), and so on, the responsibility for water management must be seriously investigated.

The River Chief System consolidates and distributes the primary responsibilities of party and government officials for water management. The distribution of responsibilities of party and government representatives for water management is generally done in three steps: first, it is about segmentation. All localities adhere to the principle of combining river basins and administrative regions, with the supra-regional rivers divided into different sections of the upper, middle and lower reaches of the rivers, and the administrative heads of each administrative region are responsible for the river sections in their jurisdictions. The second point is the clarification of the competent body, which mainly refers to the clarification and designation of the competent body of the different river sections and levels. Horizontally, according to the nature of public affairs, a general responsible person shall be designated, and vertically, according to the administrative level, different levels of responsible persons shall be designated. Third, a public notice board shall be established. The public notice board shall mainly facilitate the public to understand the main responsible persons of different departments, and provide clues for mass supervision by indicating the main overview, relevant responsibilities, management objectives, telephone numbers for supervision, etc. of the relevant area. According to the characteristics of public affairs and the characteristics of administrative division, the river chief system divides the responsibilities horizontally and vertically, and clarifies the responsible subjects of each district and level according to the three-dimensional responsibility structure, and finally realizes the blanket distribution of the responsibilities of public administration.

Thirdly, the main responsibility of Party and government officials for water management is effectively implemented. The river chief system is based on the hierarchical and hierarchical operation of the bureaucratic system, and integrates power distribution, economic incentives and internal control into the operation of the administrative contracting system, and at the same time surpasses the administrative contract system in specific operation, and is more of a bottom-up reverse operation of administrative responsibility (L. Zhou, 2022). The highest administrative responsibility hierarchy defined according to the characteristics of the characteristics of public affairs. According to the cross-administrative regional characteristics of public affairs, it is divided into several territorial administrative areas and administrative hierarchies; Designating persons responsible for the public affairs of various territorial districts and administrative hierarchies based on a hierarchical structure; the superior leads, coordinates, supervises and evaluates the subordinates, and the subordinates assume the administrative responsibilities from the bottom up, and the responsibilities of each territorial area are coordinated with each other.

Finally, the main responsibility of Party and government officials for water management should be strictly examined. The River Chief System strengthens the responsibility of public administration mainly through the evaluation system and accountability: first, the relevant tasks are integrated into the government's annual target evaluation system, and standardized and regular evaluation is carried out. Second, an accountability reward and punishment mechanism will be established. For the responsible entity that performs poorly in a series, a written report must be prepared within a certain time and corrected within a certain period; the responsible entities that fail the evaluation must be questioned and held accountable, and even the corresponding legal responsibilities should be pursued. Third, the auxiliary mechanisms for accountability assessment need to be improved. This includes the reporting system for the responsible entities, differentiated performance evaluation mechanism, lifetime accountability system, joint monitoring mechanism, etc.

To avoid unclear delineations of intergovernmental responsibilities at all levels and overlapping and confusing functions among departments at all levels. The Opinions clearly specify the organizational form of the four-tier system of river heads at the provincial, municipal, county and city levels: in each province (autonomous region or municipality directly under the central government), a general river chief shall be appointed, who shall be the chief responsible comrade of the Party committee or government; river chiefs shall be established on the major rivers and lakes in the administrative regions of each province (autonomous region or municipality directly under the central government), and the responsible comrades at the provincial level shall serve as responsible comrades; Each city, county and municipality in which rivers and lakes are located shall establish river chiefs at different levels and sections, and the responsible comrades at the same level shall serve as responsible comrades; River chiefs at the county level and above shall establish corresponding river chiefs, and the specific composition shall be determined by each locality according to the actual conditions.

The structure of the hierarchical system of contractual responsibility for local water management is based on the current system of departmental management as the basic framework. The successful implementation of the river accountability system is due to the fact that this accountability system is built on the current departmental administrative system as the basic framework, and uses the administrative contract system to allocate large and small rivers (lakes) throughout the country to party and government leaders at different levels, and through pressure-based evaluation, encourage all river accountability parties to seriously fulfill their duties and actively manage water to achieve the goal of restoring water ecology. In this system of accountability, river managers at all levels are the first in charge of the management and protection of rivers and lakes, and assume the "trinity" of responsibility for the management, governance and protection of the rivers for which they are responsible. Management means taking responsibility for the management of the river, from indirect responsibility to direct responsibility; governance means coordinating and promoting the treatment of river systems, especially the rehabilitation of pollution sources, to ensure the implementation of tasks; maintenance means coordinating and supervising the daily work of dredging, cleaning, repairing and maintaining river channels, to ensure that various long-term measures are implemented. For example, the Shanghai river chief system follows the principle of Party and government leadership and departmental linkage, and requires "establishing and improving the responsibility system with the Party and government leadership responsibility system as the core, clarifying the responsibilities of river chiefs at all levels, strengthening work measures, coordinating the forces of all parties, and forming a work pattern that covers each level and implements at all levels.

On the other hand, the construction of a hierarchical system of contractual responsibility for local water management focuses on clarifying the distribution of responsibilities between governments at all levels. The local water conservancy organization system with general river chief, deputy river chief, river chief, deputy river manager, river section chief, etc. basically follows the idea of heterogeneous responsibilities, clarifies the division of responsibilities of local governments at all levels, distinguishes the responsibilities of superiors and subordinates, realizes the correspondence between water conservancy regions and government levels, clarifies the main responsibility for water conservancy, and avoids the management of the same water area and river section at multiple levels. For example, Jiangxi Province has issued the "Jiangxi Province Implementation of the River Chief System Work Plan", which further clarifies the responsibilities of river chiefs at all levels. The top river chief and deputy top river chief of the province are mainly responsible for general supervision and planning, and lead and coordinate the protection and management of rivers and lakes in their jurisdiction as a whole (D. Liu & Richards, 2019). The river chiefs at other levels are the people directly responsible for the protection and management of rivers and lakes in their jurisdiction. They are responsible for the comprehensive prevention and control of water pollution of rivers and lakes, the ecological restoration of rivers and lakes, the correction of conspicuous problems, the inspection and cleaning of rivers and lakes, and the protection and management of rivers and lakes.

Because administrative regions and river and lake watersheds cannot completely overlap, the comments call for the city, county, and township in which each river and lake is located to appoint river chiefs at different levels and in different sections, and for the appropriate comrades at the same level to serve as river chiefs. The segmented contractual responsibility refines and concretizes the responsibilities of river chiefs and river section chiefs for water conservancy to make the river chiefs of each section be responsible for monitoring the land, take responsibility for monitoring the land, and fulfill their responsibilities to ensure the effectiveness of the River Chief System.

The segmented contract accountability system broadly clarifies accountability goals, implements stewardship responsibilities, and facilitates the development of cross-regional, networked water control through three steps: Dividing river sections, clarifying the responsibilities of each section, and clarifying who is responsible for each other's responsibilities. For example, Luojiang County in Deyang City, Sichuan Province, has renewed the responsibility mechanism of segmented management and network management, in which the Party Committee of the county manages the urban area (reservoir) and the river chiefs of cities, villages and enterprises respectively cover the water area model of the jurisdiction (Deyang Luojiang District River Chief Office, 2022). Initially, a mechanism for county-level river chiefs was established. The nine county-level rivers were divided into two or three sections, with a county-level river chief serving as the section river chief. Based on the existing liaison units of county-level river chiefs, the mechanism of responsible units for sectional river chiefs will be refined, and the responsible units will be in charge of assisting sectional river chiefs in daily inspection, management and supervision of problem elimination, and further improving the system of problem elimination and supervision; thirdly, a mechanism for Baozhen river chiefs will be established. The head of Baozhen District shall act as the head of the reservoir in the city under his responsibility, and manage and protect the small side ditches, ditches, ponds and dams and other water areas in Baozhen City.

Based on the segmented contract responsibility system, some regions have further fine-tuned the responsibilities for water conservancy and adopted the model of separate management of village and household. In recent years, Shuangfeng County in Loudi City, Hunan Province, has actively implemented the combined river purification contract system model with the help of poor households. Loudi Qingshuping Town divides the river into several river sections with the administrative village through which it flows, and 26 village-level river cleaners are contracted (J. Liu & Pan, 2018). By implementing the contract system, Qingshuping City has greatly improved local water quality and spread the experience throughout the county. Shuangfeng County, Loudi Town, has issued the Shuangfeng County River Cleaning Management Segmented Responsible Family Contracting Implementation Plan, which implements the river cleaning contracts with households and aims to maintain the ecological environment of the local watercourse through sub-village management and segmented management.

3 Politics of Diffusion of River Chief System

Since the reform and opening up, China has gradually entered a mobile society, which has a profound impact on the actual water management and its grassroots mechanisms in the new era, leading to an increasingly serious tragedy of the commons of water management, making community gradually become the consensus of water management, and the construction of ecological civilization gradually become the public will of water management under the River Chief System, thus promoting the political development of the ecological community in the new era.

Influenced by the system of household contractual responsibility for community production, the dual structure of urban and rural areas, and the policies related to urbanization, the surplus labor force has shifted from rural areas to other places, from the single-residence rural population leaving the countryside without leaving the hometown to the urban and rural population leaving the countryside and leaving the hometown. Compared with the young and vigorous migrant workers, it is mainly older people and women who stay in the countryside to work in agricultural production (Jacka, 2012). The original structure of the rural labor force has completely changed, resulting in the restructuring of water production members and changing the mechanism of water sharing.

Migration changed the membership structure of the water community. As the land allocation and land transfer system has progressed, the demand for production water has increased, especially for households. Elderly people and women living in the same area have become a new combination under the mobile society, which makes grassroots water supply and treatment difficult due to the lack of young and middle-aged workers. Due to the lack of suitable leaders, some grassroots autonomous water organizations have run into operational difficulties and are unable to perform mediation functions, which weakens the cooperation between villagers and the people. At the same time, the identity mechanism of traditional society is deconstructed, and people's interest attachment, emotional attachment and affiliation to the village are greatly reduced, which greatly weakens the enthusiasm of village groups to participate in grassroots water management, and to some extent causes the dilemma of grassroots society in water use and water management.

The intensification of mobile societies has changed the traditional mechanisms of water use. In the traditional society, the relocated farmers and low mobility in the countryside lead to a strong dependence on the land, and the local elites, internal rules, and popular customs have a strong function in mediating order and normative restraint. But increasing mobility has influenced and changed these traditions and altered the pattern of relationships between people and place. Migration encouraged land circulation, and people from other villages farming for others has become a common phenomenon in rural society. The identity of the other people outside the village challenges the internal rules and traditional customs of the village, which in turn leads to violations in water use and the broken window effect. The reason for the repeated violations by outsiders is the failure of the traditional water mediation mechanism, and villagers have little contact with the civil society of local villages, do not have to face the pressure of villagers' public opinion, and the face and moral restraint mechanism of acquaintance society works with difficulty. In addition, the state improvement of grassroots social water management facilities objectively facilitates the irrigation of farmland and reduces the need for cooperative water management by farmers to a certain extent. Therefore, the coexistence of mobile society has profoundly changed the water use pattern and management structure of grassroots society, and the question of how to adapt to the characteristics of population flows in modern society and the reality of water treatment in grassroots society and improve the quality of grassroots water management has become an important issue of the times.

In view of the numerous problems of water production in rural areas, at the beginning of the twenty-first century, the central government issued the "Guiding Opinions on Further Improving the Grassroots Water Management Service System" to improve and upgrade the grassroots water management service system in terms of administrative institutions, infrastructure, talent team, capital investment and other levels. At the same time, many localities have actively tried to introduce market mechanisms in the construction of small-scale agricultural water conservancy, such as Shandong and other localities that have carried out reform of the system of ownership of agricultural water conservancy with private investment and private operation, which has not only attracted private capital to build water conservancy, but also solved the shortage of water conservancy. In many places, water user associations have also been formed, which not only eliminate the imbalance between upstream and downstream water supply, but also avoid hitchhiking fees and save much labor in water management.

However, the market mechanism is difficult to solve the large-scale tragedy of the commons, and it is necessary to rely on the state to solve it. Although governments at all levels are aware of the principle of pollution is in the water, the root is on the shore, environmental protection does not go downstream and water conservation does not go on land has greatly limited the effectiveness of water management. Land departments focus only on groundwater, water protection departments focus only on surface water, and environmental protection departments focus only on water quality, and there is a lack of effective coordination and communication between departments, resulting in a fragmentation of water monitoring. For trans-regional rivers, local governments dealing with different levels and issues have seriously limited the effectiveness of water control due to the lack of coordination to locate water control upstream and downstream and on the left and right banks. The existence of these problems not only creates serious and unresolved environmental problems, but also poses a serious threat to economic and social development. In the long run, economic development at the expense of resource waste, pollution, and ecological damage is unsustainable, and change is inevitable.

Given the dilemma of collective action in water management, it is urgent to find new solutions to the problem. In the past, the 18th National Congress of the Communist Party of China established a series of ecological aims and approaches, such as "green water and green mountains are golden mountains and silver mountains and landscapes", and "forests, lakes and grasses are the community of life". Special governance measures are carried out, for instance, normalized centralized environmental protection inspections and three major battles, transforming people's consciousness of ecological environmental protection and marking the beginning of the era of ecological civilization. Under the guidance of the systematic water managerial community, the issue of water management has been put on the important agenda, taken into the main responsibility of local officials, and become an important matter for the environment and environmental protection department and the general public, so that systematic water management can achieve the best effect of general water control.

In the face of the worsening water management situation, old problems have not yet been solved, new problems arise from time to time, the government as the representative of the people is the "executor of the sovereign", it should act in accordance with the appeal of the "public will", comply with the public will as much as possible, and heal the water. The "Opinions on the Comprehensive Implementation of the River Control System" require local governments at all levels to carry out systematic management of all rivers in their administrative areas, and local leaders (the main Party and government leaders), as the first in charge of water control, must not only achieve the effective distribution of vertical responsibilities, but also realize the coordinated responsibility of horizontal linkage, earnestly fulfill the political, administrative, legal and professional responsibilities, and ensure the timely implementation of the public will for water control. After the official implementation of the "Opinions on the Comprehensive Implementation of the River Chief System", the river chief system was fully implemented nationwide within 19 months and achieved remarkable results.

The main reason is that under the guidance of building ecological civilization and under the requirements of "one office and two responsibilities" and "joint responsibility of party and government", local officials not only actively organize and lead the management and protection of rivers and lakes in their jurisdiction, such as water resource protection, riparian management, water pollution prevention and control, local officials not only actively organize the management and protection of rivers and lakes in their jurisdiction, such as water resources protection, riparian management, water pollution prevention and control, and supervision and evaluation of the relevant departments and river chiefs at the next level within the administrative region, but also take the initiative to coordinate the upstream and downstream and left and right banks of rivers and lakes across administrative regions to form joint prevention and control and promote the realization of the public will to build ecological civilization. At the same time, the public does not wait passively after ceding the power of consent to the state, but actively participates in various forms, such as the emergence of "village wise river chiefs", "entrepreneur river chiefs", "women river chiefs", "river chiefs of fish farmers", "red scarf river chiefs" and even "foreign river chiefs" and so on.

From the perspective of publicity for water management, the public living in the same time and space faces the same problems of water management, achieves consensus on water management, and realizes the public will of local officials to implement national water management. Driven by the intrinsic needs of water control, commonality, consensus and public opinion based on the social organizations of environmental protection, they actively participate in the construction of ecological civilization through various channels and methods, forming a joint force of the state and society to jointly control water. At the same time, with the continuous acceleration of the construction of a constitutional society in China, water control laws have been gradually introduced, which can not only ensure that local officials actively carry out their responsibilities for water control, but also promote people's conscious fulfillment of the natural obligations stipulated in the construction of ecological civilization and jointly push forward the implementation of the national public will for water control.

River Chief System is a local policy initiative in Changxing County of Zhejiang province to tackle sever blue algae pollution in 2006 in Taihu river basin, where Changxing County is located upstream. In 2008, RCS was first officially proposed in a local policy paper of Changxing County "Clear Water in Lake: a Two-Year Action Program", and "Clear Water in Lake Action" leading group and office were set up (People's Daily Online, 2008). The River Chiefs are the key leaders of the Party and government at provincial, municipal, county, township and village levels, who take charge of rivers governance and supervision of river chiefs at lower levels with their jurisdiction and adopt novel governance approaches, in order to reduce water pollution and protect water environment. RCS was firstly promoted nationwide by the General Offices of the Communist Party of China Central Committee and the State Council in the official document "Opinions on the Comprehensive Implementation of the RCS" in December 2016. Related discussions are concerned with the topics: policy remaking issue in policy transfer of RCS (Xiong, 2019; Xiong & Zhou, 2017), RCS policy implementation and diffusion path (L. Wang & Pang, 2018), success of political mobilization in RCS (Hao & Wan, 2019), political hierarchy and transboundary water governance (Chien & Hong, 2018), and river management related institution and legislation concerns (X. Yang et al., 2016).

Policy diffusion is generally termed by political scientists in a broad sense referring policy dissemination. Policy diffusion research is emphasized as the investigation of "why this happen and why governments sometimes make the same (or very similar) decisions one after another in a temporal sequence" (Knoblauch et al., 2018, p. 2). However, diffusion and transfer are often used synonymously in policy studies. It is remarked that "despite their different methodological approaches, generally identify four major mechanisms, learning, competition, coercion and mimicry, although these mechanisms receive differing degree of attention in the diffusion and transfer literatures" (Marsh & Sharman, 2009, p. 271). Along the same lines, mechanisms encouraging policy diffusion are top-down coercion, bottom-up persuasion and horizontal learning, mimicry and competition (Dobbin et al., 2007; Shipan & Volden, 2008). Marsh and Sharman (2009) assert the complimentary nature of policy transfer and policy diffusion in literature and raise five key issues of understanding policy diffusion and policy transfer. Among these, they point out that analytical focus of policy diffusion and transfer are mechanisms in diffusion and transfer, in particular, the interactions and involving of mechanisms over time. In terms of local policy innovation and dissemination, there is considerable amount of literature examining local policy innovation and patterns and mechanisms of policy diffusion. Local policy innovation is of increasing importance for local governance, and socially optimal policies diffuse in a technocratic process. The relationship between policy innovation and political opportunities are emphasized (F. Berry & Berry, 1992; F. S. Berry & Berry, 1994).

Y. Zhang and Zhu (2019) suggest noticing the gaps in studying policy diffusion theory in authoritarian context and posit "China is an excellent case for testing the generalizability of policy diffusion theory because China has a setting with a developing economy, political authoritarianism, and a Confusion culture that is radically different from typical western countries" (Y. Zhang & Zhu, 2019, p. 496). In China, policy diffusion is argued as a political process rather than a technocratic one and local officials are policy entrepreneurs in a political process (Teets, 2016). The framework of policy entrepreneur is explored in the Chinese context, demonstrating the impact of policy actors' acuity, problem definition and leading by example on policy outcomes in China (Hammond, 2013). With advanced approach of quantitative analysis, Y. Zhang and Zhu (2019, p. 509) demonstrates that "learning and imitation effects can significantly and consistently accelerate the process of innovation diffusion among provincial governments…and top-down effects can significant increase the probability of adopting new economic policy tools in China". In addition, they points out the critical role of policy environment in a transitioning country as in China owing to highly uncertain and frequently changing, suggesting that "local leaders need to be aware of various resources and opportunities to adopt an appropriate innovation" (Y. Zhang & Zhu, 2019, p. 510).

Heilmann (2008) points out the imperative role of decentralised experimentation in China's distinctive policy process remaining the prerogative of the Party leadership. This type of pilot programs, also known as point-to-surface approach, discover innovative policy instruments rather than defining policy objectives, where central policy makers encourage local officials to innovative ways of solving local problems and then experience from the pilot programs are absorbed into national policy formulation. Heilmann's argument of experimentation under hierarchy combines "decentralised experimentation with ad hoc central interference, resulting in the selective integration of local experiences into national policy-making, that is the key to understanding China's policy process" (Heilmann, 2008, p. 29). Furthermore, "the experiment under the shadow of hierarchy" framework is proposed to understand local experiments are functioned as corrective mechanisms to making and remaking the Chinese national innovation policy (Heilmann et al., 2013). According to the above discussion, political ideology, hierarchical authorities and local officials' pursuit of career promotion account for China's distinctive mode of policy experimentation and selective dissemination under hierarchy. Optimal local policy innovations and experiments in many occasions are selectively adopted or as corrective mechanism or called off in the diffusion process. This also illuminates how political resilience and adaptive governance is achieved in China (Heilmann, 2018).

Changxing County is in the northwest of Huzhou city in Zhejiang province. In 2007 the outbreak of blue algae in Taihu Lake lead to a drinking water crisis. As early as February 2008, in the revision of "Water Pollution Prevention and Control Law", "target responsibility" and "assessment and evaluation" are underlined for local governments and the party leaders in local water environment governance (X. Liu, 2017). Despite the fact that responsibilities of local officials are written in water environment related laws and regulations, implementation in the real context is lagged behind. On this basis, in 2008 RCS was initiated in the Zhejiang and Jiangsu, where local rivers were in server water pollution. Coercion, learning, competition and mimicry are considered as the key mechanisms encouraging policy diffusion (Dobbin et al., 2007). In China, local policy innovation and diffusion are posited as experimentation under hierarchy system, where cadre evaluation system works as a stimulus to policy diffusion (Heilmann, 2008, 2018; Teets, 2016). In the study of diffusion of contracting out elder care in China, Teets (2016) examines three dimensions of diffusion mechanisms which are top-down, bottom-up and horizontal mechanisms, and presents the argument of politics of diffusion in a top-down manner. These further the understanding of key diffusion mechanisms in the Chinese hierarchy context. However, these studies of policy diffusion focus on Chinese hierarchical diffusion mechanisms rather than interaction of mechanisms to produce the success or failure of diffusion in China or other developing countries. Concerning how multiple diffusion mechanisms interact and promote effective RCS policy adoption and development to match both local needs and central preferences, our following analysis examines interactions of multi-dimensional diffusion mechanisms of RCS policy development.

Changxing County has a total of 547 rivers, of which 314 cross-town rivers and 86 rivers are taken as administrative lines between townships and villages. Before RCS, the plans, standards and responsibilities of governance between villages and towns are not clear, resulting in the difficulties of implementing what is written in the "Water Pollution Prevention and Control Law". In our fieldwork, a county official in Changxing commented "*protecting rivers and lakes is a matter for the well-being of our local people. In 2003, our mayor took the lead in setting up RCS for several key rivers. The government officials and party leaders from the provincial level to village level are responsible in river governance at local levels. River chiefs are at provincial, municipal, county, town and village levels*". Why is it in 2003? Why did the mayor of Huzhou city take the lead?

Zhejiang province was one of the leading province in economic development in China. In spite of rapid economic and social development, environment issues became the major concern for the provincial leaders. Local government become the most important institutional entities in local environmental regulations (Skinner et al., 2003). Modern textile is one of the main industrial sector in Changing County, and the waste water of textile manufacturing was the key source of water pollution. In June 2003, Zhejiang province launched a "Thousand Village Demonstration and Rectification Project" under the leadership of Xi Jinping, who was the Party Secretary of Zhejiang Province. This project focuses on the improvement of the environment in rural production, rural life and rural ecology, in order to improve the rural ecological environment, the quality of farmers' life, and to balance urban and rural development (Xinhua Net, 2018). Xi Jinping's concern of environment and determination of sustainable development are highlighted in his provincial leadership in Zhejiang (Huan, 2019). In line with this project, RCS was initiated in Changxing county of Huzhou city in 2003. In our fieldwork, a meeting with Huzhou RCS Office was further explored the RCS policy initiative and innovation in Huzhou. The RCS Office officials made the comment, underlining the imperative role of good idea of political development in the initiative of RCS:

> We used to give the priority of economic development and the awareness of water protection was low. We did not realise the importance of water environment, until we read newspapers and governmental documents, also watches news on TV, knowing Xi's enormous effort and rounds and rounds fieldworks in villages in our province to emphasize the importance of environment and sustainable development. Xi's political development idea gives us a good lesson. We then had several meetings to discuss about how to tackle local river pollutions in our city.

Research interviews are also conducted with local residents in Changxing County. In the local people's opinion, the leadership demonstration is critical for the birth of various chiefs and diffusion of chief models. We are informed that in Changxing County there have been not only river chiefs but also road chiefs since 2003:"*We think the chief policy is very useful. River chiefs and road chiefs disseminate very quick from our county to other places in Zhejiang. These chiefs manage rivers and roads to keep our rivers and roads clean...In the past, no one cares about environment. The chiefs have the ability to do this and they set good examples for us.*"

When we were having a meeting in Huzhou RCS Office, we saw a statue of *Yu the Great*, who is the tamer of the flood and the founder of Xia dynasty. For 13 years, he dedicated himself to water governance project without coming home. River chiefs in Changxing county often underlined the importance of leadership demonstration in water governance, from Yu the Great to Xi's determination and leadership in Zhejiang. They believe that leadership demonstration reflects the Yu the Great's spirit, courage and determination embedded in Chinese traditional culture, not only being a water tamer but also being a political leader. This echoes with the argument that Xi's thought on eco-civilization is developed from traditional Chinese

Culture (Huan, 2019). It exerts profound impact on local and state governance in China and provides the favorable conditions of disseminating good examples, such as political demonstration and political power related resources mobilization (Hao, 2017a). Moreover, leadership demonstration also acts as a trigger for horizontal mimicry and competition.

Literature on policy diffusion in China casts light on top-down coercion mechanisms. In fact, for RCS coercion mechanisms do not play critical role in RCS diffusion. RCS is a local imitative in Changxing County, in line with the 2003 "Thousand Village Demonstration and Rectification Project" in Zhejiang under the strong political leadership of Xi Jinping, who was the Party Secretary of Zhejiang Province. Top-down diffusion of RCS is from Changxing county to its township and village levels. The total length of rivers in Changxing county is 1659 km, which has been fully managed under RCS. By 2013, Changxing County Committee and County Government have established county, township, village, and village group four levels of RCS management model, with 547 river chiefs coordinating, inspecting and supervising at various levels of rivers. At the beginning of 2014, the provincial Party Committee and the provincial Government formally established the "Five Water Co-Governance" leadership group, headed by the provincial party secretary and governor, and six deputy provincial leaders served as deputy leaders, overall and coordinated the work of water management. The leadership group office drew more than 40 'backbones' with centralised operation and eight aspects of support concerning the need of planning, funding, supervision, assessment, science and technology support, regulations and executing command. In addition, 30 inspection groups were set up in the provincial government to conduct in-depth inspection of implementation of water management. Responsibilities of river chiefs are shouldered at different administration levels (Xu & Lu, 2019).

Meanwhile, horizontal diffusion of RCS goes from Changxing County to other counties in Huzhou city and cities along the Taihu Basin in China. As early as in June 2008, Changxing's neighbor Jiangsu province issued the "Notice of the General Office of the People's Government of Jiangsu Province on the Implementation of the Dual River Chief System in the Main Rivers into Taihu Lake". The Jiangsu Provincial People's Government decided to implement innovative "Dual RCS" for 15 rivers into the Taihu Lake. Two types of river chiefs are defined: provincial river chiefs and local river chiefs. Provincial river chiefs are led by the provincial government, who are from the provincial Taihu Water Pollution Prevention and Control Committee and the provincial Bureaus. Local river chiefs are city, county and district People's Government officials. The main responsibility of river chiefs in Zhejiang and Jiangsu are in common, organizing, leading and planning the implementation of water environment management, coordinating the resolution of disputes and problems at work, taking responsibilities of supervising and inspecting, as well as looking after project planning and funding issues (M. Zhu, 2017).

Supervision is one of rules of institutional design to ensure successful self-governance in common pool resources management (Ostrom, 1990). Public supervision is one of crucial mechanisms to make RCS operate and diffuse effectively from Changxing to Zhejiang and neighboring local areas. It provides an effective channel to respond local governance problems (Hao, 2019). Furthermore, we argue that governing networks based on pubic supervision and participation in process RCS development provide major impetus for bottom-up diffusion of RCS. Since 2003 the initiative of RCS in Changxing, public participation has become more and more active with diversified forms and characteristics. On the volunteering basis, there are groups of women river chiefs, little young pioneer river chiefs, youth river chiefs, elder river chiefs. In addition to the groups of volunteering river chiefs, there are active local people who are informally participating in looking after local rivers. Local water governing networks also sees the significant role of informal but solidary groups in public goods provision (Tsai, 2007a).

During our fieldwork, we are told by the local volunteers and groups of volunteering river chiefs that "*following the RCS billboards, we keep an eye not only on what are in rivers but also the job by the river chiefs whose names, responsibilities, phone numbers and their bosses' numbers are displayed on the billboards. When there are problems, we just make phone calls to raise problems…We propose potential solutions. Our volunteering river chiefs are useful assistants to the official river chiefs to identify local problems and know local people's need in the local river governance.*"

It is argued that local need incentivizes policy experimentation. Local officials' behavior can be explained by local social networks and policy innovation is influenced by how local officials respond to local problems, being held accountable for local water environment (Teets, 2016; Tsai, 2007b). In the case of RCS, river chiefs are under the supervision and receiving persuasion from local water governing networks that is composed of active volunteering river chiefs and related groups in Changxing. Vertically, various forms public participation in the governing networks promote the bottom-up diffusion of informal river chiefs groups and informal institutions in water governance through supervision and persuasion. This further incentivizes RCS policy experimentation and effective horizontal diffusion.

Effective horizontal diffusion mechanisms are explained by leaning, mimicry and competition (Marsh & Evans, 2012; Teets, 2016; Y. Zhang & Zhu, 2019). Learning among provinces is a main approach for RCS diffusion between Zhejiang and Jiangsu provinces. Two adjacent provinces facing the common challenge of Taihu Lake pollution are closely interactive in mutual learning, mimicry and competition before issuing the official document "Opinions on promoting the RCS" by the General Offices of the Communist Party of China Central Committee and the State Council in December 2016. In the case of RCS diffusion, horizontal diffusion mechanisms are mainly important to cities and provinces affected by the pollution in Taihu Basin, such as Zhejiang, Jiangsu, Shanghai, Fujian, and south of Anhui. Local officials in Changxing told us "*other provinces learn RCS from us…we felt the pressure from Jiangsu and Shanghai in the competition of water governance. We think all kinds of best solutions to improve our RCS work, and also invest in high technology, such as robotic equipment for cleaning water pipes and smart phone device system to track river chiefs' patrol along and with boats on rivers*".

In 2017 and 2018, we conducted fieldworks in S county of Anhui province who is adjacent to Zhejiang province along the Taihu Basin. In Anhui, RCS is basically learned and adopted in S county, and the diffusion does not reach to the provincial level. We were informed that Taihu Basin in Anhui only a small section in S county and pollution is comparatively low there owing to no industrial pollution in south of Anhui. Interestingly, we were also informed by the county river chief that "*our county is relatively poor compared with our neighbor cities in Zhejiang and Jiangsu. We have to be strategic when adopt a policy according to our real situation. It is not sensible to copy everything from them. We learn from their advanced experience but selectively... We listen to the central command. Our president promotes 'Green Water and Green Mountain' last year. We speed up to learn and copy the experience from Zhejiang, making river chiefs billboards and holding river chiefs meetings more frequently...As a matter of fact, water quality in our country is much better than Zhejiang. Considering the economic resources and pollution conditions, so we selectively use some of good experience from other provinces*".

In the horizontal diffusion, diffusion mechanisms are selectively and effectively applied according to local water pollution and economic conditions. From the horizontal diffusion within Zhejiang province, we understand that RCS diffuses actively and horizontally from one city to another. Despite water pollution at various levels, above analysis explains that coercion with the promotion of good idea of political development and leadership demonstration give the impetus for horizontal competition. Besides, economic and pollution conditions affect the strategies of horizontal diffusion of RCS. In literature, purpose of mimicry is argued not being entirely for policy outcome to solve practical problems in many situations (Shipan & Volden, 2008; Smith, 2013). RCS's success diffusion makes a light shine on politics of diffusion and its mechanisms of matching local needs, central preferences and local policy environment.

Above analysis illuminates in RCS diffusion how multiple diffusion mechanisms interact and promote effective policy adoption to match local needs, central preferences and local policy environment. The following discussion focuses on the mechanisms that underpins the success of RCS diffusion to match local needs and central preferences in the Chinese 'politics of diffusion' mode. Studies of local policy innovation and diffusion in China assert the importance of cadre responsibility system, and local governors expect future career promotion by being good policy entrepreneurs and implementers (Edin, 2003). In the Chinese cadre evaluation system, meeting local needs, as one of the evaluation criteria, has fundamental impact on local development (Heberer & Trappel, 2013). Arguably, the cadre responsibility system is crucial to good performance in policy innovation and implementation. As it discusses above, the study finds the variations of policy implementation and selective policy adoption strategy in S county of Anhui province, who is confined by the local economic and environmental conditions. At this point, the cadre responsibility system gives inadequate explanation. Skinner et al. (2003) study the changing role of local government in social and environmental regulation in Huzhou Municipality where Changxing county is located. They find the ability of local government's selective policy implementation conditioned by the local priorities, and point out it is because of administrative decentralization and devolution of administrative responsibility. Furthering the changing role of local government in front of the trade-off between local economic development and environmental regulation, RCS reinforces the political responsibility of the river chiefs, key leaders of the Party and government at provincial, municipal, county, township and village levels (Hao, 2017a).

Although it is stated that China's politics of diffusion is demonstrated by top-down mechanisms (Teets, 2016), our study contends that bottom-up diffusion mechanisms are critical to effective policy diffusion in.

China's RCS 'politics of diffusion' mode (Fig. 1). Modification of diffusion mechanisms are generated to facilitate the RCS diffusion in different localities.

This mode sheds light on the key political factors driving the success of RCS diffusion from Changxing county to the nation: political responsibility, political responsiveness and political opportunity. Under the Party and Government leadership, top-down diffusion works with the coercion mechanism propelled by the good idea of sustainability in political development and political leadership demonstration. This is resonated by the status of Yu the Great, and Xi's significant effort to rural development and environmental governance from the Zhejiang policy experiment to the promulgation and diffusion of RCS in China. Under the pressure of coercion mechanism, learning, mimicry and competition mechanisms are strategically applied by local actors in local policy environment. Interaction of coercion and learning, mimicry and competition mechanisms promote effective policy adoption to match both local needs, central preferences and local policy environment. Bottom-up diffusion mechanisms are underlined by public participation and persuasion in the process of RCS diffusion. The importance of public participation is addressed in sustainable river governance in China (X. Yang et al., 2016). Over the course of RCS diffusion, the pressure of participation and persuasion mechanisms nudge the political responsiveness from the top to public interests. Absorbing and adapting mechanisms of diffusion are produced as the modification of diffusion mechanisms by interaction of mechanisms in top-down, bottom-up and horizontal dimensions. These two diffusion mechanisms enhance the actors' responsiveness to water environmental issues, related policies implementation and collective action in specific policy environment.

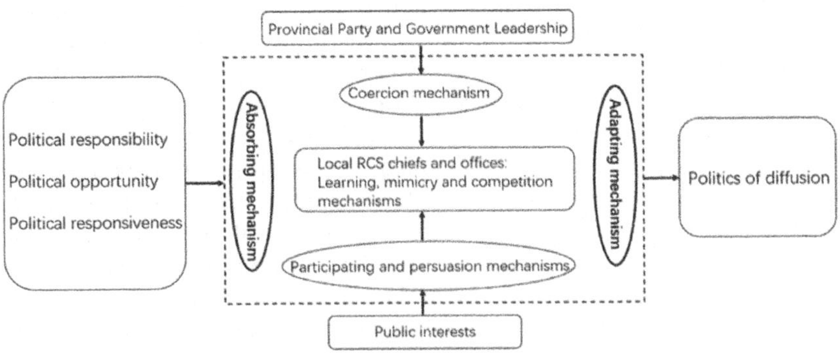

Fig. 1 China's RCS 'politics of diffusion' mode

China's RCS 'politics of diffusion' mode also emphasizes political opportunity as one of the political drivers for the success of RCS diffusion. The importance of grasping political opportunity is illustrated in the interaction between horizontal diffusion mechanisms and top-down diffusion mechanisms. In Zhejiang, leadership demonstration is embedded in coercion mechanism, and it made local officials in Changxing and neighboring cities identify and grasp the political opportunity to trigger RCS diffuse swiftly in Zhejiang and Jiangsu provinces in the early 2000s. The above analysis shows that S county in Anhui province demonstrates that grasping the political opportunity is conditioned by the local economic and environmental context. This casts light on the importance of local policy environment in understanding multiple diffusion mechanisms (Y. Zhang & Zhu, 2019).

China's 'politics of diffusion' mode illuminates the interaction of diffusion mechanisms. In the political process of diffusion, two modification diffusion mechanisms are produced, absorbing and adapting mechanisms, as presented above. The following analysis concerns how politics of diffusion mechanisms work in RCS to influence the outcome of diffusion and local policy adoption. Based on study of RCS in Zhejiang and Anhui provinces, we compare and identify two types modification diffusion mechanisms, and their effect on the outcome of diffusion. Over the course of interaction of diffusion mechanisms in top-down, bottom-up and horizontal dimensions, RCS diffusion witnesses the modification of diffusion mechanisms within a particular local policy environment. This has significant impact on local policy adoption. Changxing County is the birthplace of RCS when confronting severe water pollution in the Taihu Basin in the early 2000s. Encouraged by the provincial leadership demonstration and good idea of sustainability in political development, party leaders from the provincial to local levels shoulder the responsibility and seize the political opportunity in environmental governance. Political responsibility in water governance was clearly established as the top priority for party leaders in Zhejiang as early as 2003. In an interview, a provincial river chief in Zhejiang explained how the political responsibility was established and RCS was originated in Zhejiang: "*the idea of water governance should date back to 2003. Xi's strategic thought on ecology and environment exerts profound impact on our governors, when he was our provincial leader. This lays solid foundation of institutional design of RCS in Zhejiang. In RCS, political responsibility is the top priority and this I why the general RCS office is established in the Provincial Government rather than the Provincial Water Resources Bureau.*"

Shouldering the political responsibility, river chiefs at the local level consider political opportunity as leadership demonstration by good performance for local people's livelihood improvement. A county river chief in Changxing County in the municipality of Huzhou told us that "*political responsibility gives us the opportunity to being a good local leader for our local people*". Huzhou Municipality was rewarded the *Yu the Great Award* for three successive years. This award is the most prestigious thing for water governance in Zhejiang province. The county river chief in Changxing further explained the reason of the reward: "*our leadership is demonstrated by our ability to motivate local people's participation. There are more than thousands of volunteers participating in monitoring river surface and reporting water pollution in every corner of Huzhou. Honestly, this gives our river chiefs pressure, but it is good pressure for our chiefs' work. Moreover, with the advance of technology, people's participation makes us know water pollutions and related problems in a flash. It helps us to deal with water problems in time*". Local people's participation in water governance gives the pressure on river chiefs' political responsibility and nudges the political responsiveness from the river chiefs in the locality. Apparently, RCS diffusion from Changxing County to Zhejiang province uses top-down coercion mechanism, horizontal mimicry, learning and competition mechanisms, and bottom-up participating and persuasion mechanisms. Modification mechanisms are identified as adapting and absorbing mechanisms. Table 1 illustrates how politics of diffusion mechanisms work in RCS, and influence the outcome of diffusion and local policy adoption in Zhejiang. In RCS diffusion, besides top-down, bottom-up and horizontal diffusion mechanisms, we discover modification mechanisms that accelerate the successful diffusion of RCS form a place to the whole Zhejiang province. Adapting mechanism is generated when diffusion adapts local policy environment; and absorbing mechanism is generated to facilitate the RCS diffusion by matching the public interests with the central preference expressed in the official RCS promotion documents. The outcome of diffusion is illuminated by how RCS is adopted locally. In Zhejiang, RCS is not only diffused and implemented, but also policy innovation is carried out in various localities to adapt local policy environment. A provincial river leader commented: "*in order to avoid RCS becomes a mere formality, we make each river an archive and 'one river one policy' including every details of the river pollution points, water quality and open sewers. When the water pollution point is identified, we make 'one point one policy' at the local level*". In line with the diffusion of RCS in Zhejiang, environmental related chiefs such as forest chiefs and road chiefs are institutionalised from Changxing to neighboring counties, cities and provinces. The modification of policy diffusion mechanisms, adapting and absorbing mechanisms, enhance the actors' responsiveness to policy implementation and innovation in particular local policy environment.

Table 1 Diffusion mechanisms and modification type I

	Political factors	Actors	Diffusion mechanisms	Modification of mechanisms	Diffusion outcome
Local policy environment	Political responsibility ++	Provincial river chiefs ✔	Coercion	Adapting & Absorbing	Implementation & Innovation
	Political opportunity ++	Local river chiefs ✔	Mimicry learning competition		
	Political responsiveness ++	The public ✔	Participating persuasion		

For S county of Anhui province, water resources are abundant and in a fair condition. When RCS was promulgated at the end of 2016, river chief offices have been established from provincial to local levels in Anhui, and RCS was implemented in S county. By comparing the diffusion mechanisms with that in Zhejiang, it reveals a variation in diffusion mechanisms and outcome of diffusion (Table 2). Little public involvement is associated with low level of political responsiveness. Political responsibility for river chiefs in S county is given by the top levels of river chief offices. In terms of RCS diffusion, besides top-down coercion mechanism, the horizontal dimension is mainly mimicry mechanism and bottom-up mechanism is apparently very weak. Zhejiang's experience of RCS is selectively copied in S county to adapt local policy environment.

In addition to top-down, bottom-up and horizontal diffusion mechanisms, adapting mechanism in policy diffusion leads to selective RCS implementation in S county. In the case of RCS diffusion from Changxing county to Zhejiang province, bottom-up mechanisms play critical role in China's politics of diffusion. In the process of policy diffusion, official river chiefs' political responsiveness is activated by public participation and persuasion. Bottom-up diffusion mechanisms provide fresh impetus for river chiefs' shouldering political responsibility and grasping political opportunity to achieve effective RCS diffusion. Indeed, diffusion and modification mechanisms in S county makes RCS implemented adaptively and selectively in its local context. In the politics of diffusion mode, exercising political responsibility is the priority for S county, which is highlighted in the "Opinions on the Comprehensive Implementation of the RCS". Political opportunity and responsiveness are considered as the "priority in the second place" by the river chiefs in S county, which is explained by one of the river chiefs: "*We learn and copy good experience from Zhejiang. The priority is to follow what is stressed in the document from the top. The emphasis is our [river chiefs] political responsibility in implementing RCS. Step by step, we will consider and organize activities for local people's participation in monitoring water quality and telling us their opinion about our work. I believe this will improve our work performance and hopefully our work will be recognized by our bosses in the future.*"

Table 2 Diffusion mechanisms and modification type II

	Political factors	Actors	Diffusion mechanisms	Modification of mechanisms	Diffusion outcome
Local policy environment	Political responsibility ++	Provincial river chiefs ✔	Coercion	Adapting	Implementation
	Political opportunity +	Local river chiefs ✔	Mimicry		
	Political responsiveness	The public x	x		

The success of RCS diffusion is apparently shown in the outcome of improving river environment in China. Under the leadership of Party and Government at each level, water pollution in Taihu Basin is tackled with the achievement of significant results. Following Zhejiang, Jiangsu and Shanghai, Fujian, Beijing and many other provinces began to implement the RCS. In December 2016, Central Office of the Communist Party of China and the General Office of the State Council issued "Opinions on the Comprehensive Implementation of the RCS". On 1 January 2017, General Secretary Xi Jinping highlighted in his New Year's message that every river should have a river chief. In June 2017, the RCS is firstly written in the newly amended "Water Pollution Prevention and Control Law of the People's Republic of China". By the end of June 2018, 31 provinces, autonomous regions and municipalities directly under the Central Government have fully established the RCS.

Political factors in the politics of diffusion mode are identified as political responsibility, political responsiveness and political opportunity. In previous studies, politics of diffusion is considered as being dominated by top-down diffusion mechanisms (Teets, 2016). Our study finds that bottom-up diffusion mechanisms play important role in China's politics of diffusion mode. The bottom-up diffusion mechanisms give impetus to modification of mechanisms and nudge the political factors towards influencing effective diffusion. Modification of diffusion mechanisms are created in accordance with particular local policy environment, promoting effective policy adoption to match both local needs and central preferences. RCS propels political responsibility and political responsiveness to the needs of local river governance within specific local policy environment. Interactions of multi-dimensional mechanisms accelerates the diffusion from a local initiative to a nationwide adoption to grasp political opportunity. It also demonstrates the political resiliency and adaptability in China's politics of diffusion, which account for the success of China's adaptive local governance and policy innovation through policy experimentation from local experimentation points to national policy (Heilmann, 2018). Nonetheless, it warrants attention that diffusion mechanisms and different approaches are conditioned by local policy environment, resources limit and political priorities. In order to achieve promising local policy adoption and be environmentally sustainable, it raises the importance of encouraging public participation in policy diffusion process to form bottom-up diffusion mechanisms which nudge effective policy diffusion.

References

Agrawal, A., & Ostrom, E. (2001). Collective action, property rights, and decentralization in resource use in India and Nepal. *Politics & Society, 29*(4), 485–514. https://doi.org/10.1177/0032329201029004002

Berry, F., & Berry, W. (1992). Tax innovation in the states—capitalizing on political opportunity. *American Journal of Political Science, 36*(3), 715–742. https://doi.org/10.2307/2111588

Berry, F. S., & Berry, W. D. (1994). The politics of tax increases in the states. *American Journal of Political Science, 38*(3), 855–859. https://doi.org/10.2307/2111610

Chien, S.-S., & Hong, D.-L. (2018). River leaders in China: Party-state hierarchy and transboundary governance. *Political Geography, 62*, 58–67. https://doi.org/10.1016/j.polgeo.2017.10.001

Deyang Luojiang District River Chief Office. (2022). *Comprehensively strengthen the relevant content of the main points of the river and lake chief system.* District Water Bureau. https://www.luojiang.gov.cn/gk/xwzx/bmdt/1523933.htm

Dobbin, F., Simmons, B., & Garrett, G. (2007). The global diffusion of public policies: Social construction, coercion, competition, or learning? *Annual Review of Sociology, 33*, 449–472.

Edin, M. (2003). Remaking the Communist Party-State: The cadre responsibility system at the local level in China. *China: An International Journal, 1*, 1–15. https://doi.org/10.1353/chn.2005.0013

Hammond, D. R. (2013). Policy entrepreneurship in China's response to urban poverty. *Policy Studies Journal, 41*(1), 119–146. https://doi.org/10.1111/psj.12005

Han, Z., & Li, C. (2021). How responsibility is constructed: River chief system and its implementation in S City. *Public Management Research (in Chinese), 44*(1), 137–145.

Hao, Y. (2017a). Establishing responsibility of local chiefs in the "river chief system" context. *Journal of Henan Normal University (Philosophy and Social Sciences Edition) [in Chinese], 44*(5), 13–18 [in Chinese].

Hao, Y. (2017b). The responsibility of local officials in the river chief system. *Journal of Henan Normal University (Philosophy and Social Sciences Edition) (in Chinese), 44*(5), 13–18.

Hao, Y. (2019). Public responsibility: Historical logic of the creation and development of river chief system. *Journal of Yunnan Social Science,* (4), 60–66 [in Chinese].

Hao, Y., & Wan, T. (2019). Consensus mobilization: Frame analysis of river chief system activating public responsibility. *Journal of Guangxi University (Philosophy and Social Sciences Edition), 41*(04), 133–140 [in Chinese].

He, Y. (2019). The logic of "five-in-one" and "four comprehensive strategies". *People's Forum (in Chinese),* (15), 15–17.

Heberer, T., & Trappel, R. (2013). Evaluation processes, local cadres' behaviour and local development processes. *Journal of Contemporary China, 22*(84), 1048–1066. https://doi.org/10.1080/10670564.2013.795315

Heilmann, S. (2008). From local experiments to national policy: The origins of China's distinctive policy process. *The China Journal,* (59), 1–30. https://doi.org/10.2307/20066378

Heilmann, S. (2018). *Red Swan: How unorthodox policy-making facilitated China's rise.* Chinese University Press.

Heilmann, S., Shih, L., & Hofem, A. (2013). National planning and local technology zones: Experimental governance in China's torch programme. *The China Quarterly, 216*, 896–919. https://doi.org/10.1017/S0305741013001057

Huan, Q. (2019). The factors of traditional culture in Xi Jinping's thought on eco-civilization. *Journal of Fujian Normal University (Philosophy and Social Sciences Edition),* (6), 1–9 [in Chinese].

Jacka, T. (2012). Migration, householding and the well-being of left-behind women in rural Ningxia. *The China Journal,* (67), 1–22.

Knoblauch, D., Mederake, L., & Stein, U. (2018). Developing countries in the lead—What drives the diffusion of plastic bag policies? *Sustainability, 10*(6), 1994.

Li, W., Liu, J., & Li, D. J. J. o. e. m. (2012). Getting their voices heard: Three cases of public participation in environmental protection in China. *Journal of Environmental Management, 98*(15), 65–72.

Linde, J., & Peters, Y. J. P. P. (2020). Responsiveness, support, and responsibility: How democratic responsiveness facilitates responsible government. *Party Politics, 26*(3), 291–304.

Liu, X. (2017). Analysis of basic structure and key issues of the implementation of river chief system. *Water Resources Development Research,* (11), 25–27 [in Chinese].

Liu, J., & Pan, L. (2018). Loudi pilot river cleaning management subcontracted to households. http://news.ldnews.cn/news/loudi/201812/589227.shtml

Liu, D., & Richards, K. (2019). The He-Zhang (River chief/keeper) system: An innovation in China's water governance and management. *International Journal of River Basin Management, 17*(2), 263–270.

Marsh, D., & Evans, M. (2012). Policy transfer: Coming of age and learning from the experience. *Policy Studies, 33*(6), 477–481. https://doi.org/10.1080/01442872.2012.736795

Marsh, D., & Sharman, J. C. (2009). Policy diffusion and policy transfer. *Policy Studies, 30*(3), 269–288. https://doi.org/10.1080/01442870902863851

Ostrom, E. (1990). *Governing the commons: The evolution of institutions for collective action.* Cambridge University Press.

People's Daily Online. (2008). *"Clear Water into Lake" changing county of Zhejiang Province moves to change the water quality of Taihu Lake.* https://doi.org/www.qingyuan.people.com.cn

Perdue, P. C. (2013). Ecologies of empire: From Qing Cosmopolitanism to modern nationalism. *Cross-Currents: East Asian History and Culture Review,* (8), 4–30.

Rawls, J. (2020). *A theory of justice: Revised edition*: Harvard university press.

Shipan, C. R., & Volden, C. (2008). The mechanisms of policy diffusion. *American Journal of Political Science, 52*(4), 840–857.

Skinner, M. W., Joseph, A. E., & Kuhn, R. G. (2003). Social and environmental regulation in rural China: Bringing the changing role of local government into focus. *Geoforum, 34*(2), 267–281. https://doi.org/10.1016/S0016-7185(02)00077-5

Smith, M. P. (2013). The global diffusion of public policy: Power structures and democratic accountability. *Territory, Politics, Governance, 1*(2), 118–131. https://doi.org/10.1080/21622671.2013.808162

Teets, J. (2016). Policy diffusion in China: Contracting out elder care. *China: An International Journal, 14*(3), 88.

Tsai, L. L. (2007a). *Accountability without democracy: Solidary groups and public goods provision in rural China.* Cambridge University Press.

Tsai, L. L. (2007b). Solidary groups, informal accountability, and local public goods provision in rural China. *The American Political Science Review, 101*(2), 355–372.

Wang, S., & Yao, Y. (2007). Grassroots democracy and local governance: Evidence from rural China. *World Development, 35*(10), 1635–1649. https://doi.org/10.1016/j.worlddev.2006.10.014

Wang, L., & Pang, R. (2018). Temporal and Spatial evolution mechanism and policy diffusion path of China's public policy: A study of the implementation and changes of river-chief system. *Chinese Public Administration, 395*(5), 63–69 [in Chinese].

Xinhua Net. (2018). Painting a new picture of the beautiful countryside in New Era: General secretary Xi Jinping and Promoting Zhejiang's "thousand village demonstration, thousand village rectification" Project. http://www.xinhuanet.com/politics/2018-04/23/c_1122728788.htm

Xiong, Y. (2019). Research on "policy re-construction" in regional policy transfer in China: Based on grounded theory analysis of the river-chief system transfer in a city of Jiangsu Province *Journal of Public Management, 16*(3), 131–144 [in Chinese].

Xiong, Y., & Zhou, J. (2017). Policy reproduction in "river-chief system" transfer: Influencing factors and model generalization—Based on the qualitative comparison and analysis of multiple cases in Jiangsu Province. *Journal of Gansu Institute of Public Administration,* (1), 37–47 [in Chinese].

Xu, G., & Lu, Y. (2019). On the governance experience and future of "five water co-governance" from the perspective of collaborative governance theory. *Administration Management Reform [in Chinese],* (2), 33–40 [in Chinese].

Yang, X., Lu, X., & Ran, L. (2016). Sustaining China's large rivers: River development policy, impacts, institutional issues and strategies for future improvement. *Geoforum, 69*, 1–4. https://doi.org/10.1016/j.geoforum.2015.11.019

Zhang, Y., & Zhu, X. (2019). Multiple mechanisms of policy diffusion in China. *Public Management Review, 21*(4), 495–514. https://doi.org/10.1080/14719037.2018.1497695

Zhou, L. (2022). The administrative contracting system and Chinese state capacity. *Open times (in Chinese), 44*(4), 28–50.

Zhu, M. (2017). A decade of review and prospect of Taihu Basin governance. *Environment Protection,* (24), 34–38 [in Chinese].

Chapter 3
Bottom-Up Mechanisms: Civil River Chiefs

1 The Role of Civil River Chiefs

In addition to the official river chiefs established by the state, local governments, especially at the county level and below, actively enlist rural villagers, media workers, enthusiastic entrepreneurs, social volunteers, and other groups to participate in river management and protection, and establish various civil river chiefs. There are village river chiefs, enterprise river chiefs, women river chiefs, cleaner river chiefs, postmen river chiefs, and so on, to address the multiple roles of civilian river chiefs in managing rivers.

First, oversight of civilian river officers has revealed problems. All types of civilian river supervisors use their proximity to river sections, familiarity with water conditions, and flexible time to actively participate in monitoring and protecting river sections, assist government river chiefs in uncovering river management problems, and monitor the government's work in river management. The River Supervision Bureau of Muyang County in Jiangsu Province has invited county People's Congress deputies, CPPCC members, enterprise representatives, lawyers, media workers, retirees, social activists and other representatives from all walks of life to serve as social river chiefs and enterprise river chiefs. The main duties of social river chiefs and corporate river chiefs are to report illegal acts to pollute the water environment and damage to the water environment, collect and understand the suggestions and opinions of the public on the work of the river chief system, monitor and evaluate the impact of river (section) management and administration in supervision regularly every month, and report relevant situations such as sewage discharge, garbage dumping, reclamation and destruction of administrative facilities found in supervision and inspection without delay.

Secondly, civil river chiefs are involved in solving problems. Civil river chiefs belonging to different industries have various characteristic social resources and can link them together, and with the blessing of their own professional advantages and local knowledge, they are an important force to solve open problems in river management and protection. Local governments at all levels actively involve all kinds

Y. Hao and T. Wan, *The River Chief System and An Ecological Initiative for Public Participation in China*, https://doi.org/10.1007/978-981-99-4921-2_3

of civilian river officers in solving local water management problems. In Zhongguan City, Deqing County, Zhejiang Province, eight civilian river chiefs have been hired to participate in river management and protection. Wu Jianrong, a villager who has been involved in aquaculture for more than 20 years, knew the importance of good water quality and immediately took the initiative to run for office after learning about the recruitment of civilian river chiefs. After his successful candidacy for the post of civil river superintendent, Wu Jianrong inspected the river section he was in charge of every day and frequently visited the surrounding aquaculture farms to publicize the benefits of underwater treatment and mobilize the surrounding aquaculture households to jointly improve water quality.

Third, civil river chiefs' participation sets role models and facilitates educating activities. The fundamental solution for long-term water management is to create an image throughout society that is appreciated and shared by all. The importance of social subjects for water management at the ideological level is the prerequisite for their participation in water management, and the creation of a good social climate for water management is a necessary condition for the formation of social subjects' awareness of water management. The creation of this good social climate is inseparable from government initiative and active public support. In this regard, local governments at all levels have taken the initiative to hire private river chiefs to help publicize relevant laws, regulations, policies and environmental protection knowledge related to the river chief system in order to promote the river chief system and achieve greater results. For example, in Changshan County, Quzhou City, Zhejiang Province, the cyclists forming River Chiefs Alliance was established with 40 cyclists as river chief teams including 2,000 participants, and the path of Cycling River Chiefs covers 180 administrative villages in the county, widely spreading the concept of non-profit water governance.

2 Women's Participation and Cross-Basin Collective Action

From 2018 to 2022, we have conducted several fieldworks in Zhejiang, Anhui and Hubei provinces. River Chief System was initiated in Changxing county of Huzhou city in Zhejiang province in 2006 to tackle blue algae pollution in Taihu river basin crossing both Zhejiang and Jiangsu provinces. In the west of Zhejiang province, the Xin'an River, crosses the provinces of Zhejiang and Anhui, is the common mother river of She County in Anhui Province and Chun'an County in Zhejiang. In order to protect the Xin'an River Basin, the Women's Associations of Zhejiang and Anhui have developed joint river management plans.

One of the authors had the opportunity to interview Mei a member of the She County Women's Association in Huangshan city of Anhui province. Huangshan city took the lead in introducing the river chief system in the province in 2013 to improve the water quality of the Xin'an River Basin. The daily work of the women river chief is to inspect the river, mainly to find out if there are sewage discharges and waste, and report unsolved problems to the governmental river chiefs. Mei commented her motivation in participation as a river chief working with colleagues from Zhejiang: "the protection of river and environment is the responsibility of each of us, and I would like to use the knowledge I have learned today to be a good order keeper, informant and liaison in our community, and implement the concept of loving the river and protecting the river, to arouse the public's awareness and their participation in protecting water environment". The County Women's Associations organized river inspecting teams during the week and various events at weekends. Women river chiefs distribute brochures and to the public to spread knowledge about the river and lake chief system, and water pollution prevention and control. Women river patrol teams, consisting of a group of three to five women river chiefs, walked all the way to pick up leaves, cigarette butts and other garbage along the banks of the river. They walked about 4 km every day to inspect and solve all kinds of problems related to the river environment.

Mei is seen as "iron lady" in local people's eyes, with formal work in the county women's association and informal work as a river chief. She took her son and friends together with her at weekend to clean the bank of Xin'an river, and believed her action of water protection had influenced many people in her family and community. Importantly, Mei emphasized being a river chief is proud, and her action was influential because it is a governmental promoted and prioritized water governance action. This confirms that in China women's empowerment cannot be achieved without the efforts of the government (L. Yang, 2012). Mei told us that her sense of pride is not only from the environmental impact of her role in river governance but also from the recognition by the local government praising her with award certificates for her as "lady warrior of river governance".

3 Public Participation Promoting Environmental Well-Being

As part of the comprehensive implementation of the river chief system across the country, the public can provide bottom-up feedback on water control issues through multiple monitoring channels of formal and informal institutionalization and influence higher levels of government power to promote effective monitoring (Stromseth, 2017). The implementation of the volunteer project "Red Scarf Little River and Lake Chief" is an innovative approach to promote the construction of the new ecological city in Wuhan city of Hubei province. It is hoped that the young pioneers will actively participate in the volunteer service activities of river and lake governance. This has attracted students from primary, secondary and high schools to participate in river and lake protection. At weekends students distributed leaflets and environmental materials to passing tourists to ask tourists not to throw plastic bags into the Han River. Under the premise of ensuring safety, several primary students from the garbage collection team accompanied by governmental river chiefs and volunteers to collect all kinds of garbage along the river bank.

The retired cadres have been actively river chiefs, who are named silver-haired river chiefs. The silver-haired river chiefs who are river patrol volunteers were enthusiastic, united and cooperative, unafraid of dirt, toil and fatigue, picking up tongs, garbage bags and brooms to collect plastic bags, plastic bottles, paper scraps and other household garbage along the river, and also cleaned up the riverbank and both sides of recreational and entertainment facilities. The silver-haired river chiefs became ambassadors of the river basin management system, distributing the River Chief System Leaflet to the people in the area, and actively promoting the beneficial measures of the river basin management system. This made the work of the river basin management system deeply rooted in people's hearts, and created a good social atmosphere of believing "everyone is a river chief, and everyone protects the river".

Based on the service center for retired cadres in each district, retired cadres and Party members to give full use of their political advantages, experience, prestige and professional expertise, and actively participate in river inspection and river protection activities. They guide the silver-haired river chief to be a good inspector who actively discourages all kinds of uncivilized behavior, such as discharging sewage into rivers and lakes, and dumping garbage and waste. The aim of silver-haired river chiefs' action is the same, to be a good propagandist and widely publicizing Xi Jinping Thought on ecological civilization and the laws and policies of river and lake protection, and also play a role model for the public to participate in the protection of rivers and lakes.

During our fieldworks in Anhui and Hubei provinces we talked with some silver-haired river chiefs about their roles in river and lake protection. The most common and impressive comments are silver-haired river chiefs believe they have profound experiences compared with young river chief and are more familiar with the environment conditions, which are useful to make suggestions and proposals on local river and lake management and protection. Their common practices are being good liaisons, and collecting feedback from local people about the work of the river and lake chiefs. They described that they act as a bridge between the government and the local people.

References

Stromseth, J. (2017). *China's governance puzzle: Enabling transparency and participation in a single-party state*. Cambridge University Press.

Yang, L. (2012). Empowered or disempowered? Women's participation in a development project in rural China. *Asian Journal of Women's Studies, 18*(3), 38–67. https://doi.org/10.1080/12259276.2012.11666131

Chapter 4
From Governmental River Chiefs to Public Participation

1 Rivers Chiefs' Action

In the management and improvement of the quality of rivers and lakes, the action of river patrols by the river chiefs of each river section is an important means of dynamically assessing the quality of rivers and resolving problems of river management and protection in a timely manner. Depending on the administrative level to which the river section belongs, the number of river patrols per year also varies. The "Guidelines of the Ministry of Water Resources for Further Strengthening the Fulfilment of the Duties and Responsibilities of River and Lake Chiefs" stipulates that the river chief as a whole shall conduct river patrols not less than once a year, the provincial river and lake chief shall conduct river patrols not less than twice a year, the municipal river and lake chief shall conduct river patrols not less than three times a year (at least once every six months), the county-level river and lake chief not less than once a quarter, the township-level river and lake chief not less than once a month, and the village-level river and lake chief not less than once a week (General Office of the Ministry of Water Resources, 2019). The "Anxi County River Chief and Lake Chief Performance Regulations (Trial)" of Fujian Province stipulates the following: The county-level river chief and lake chief must be present at least once a month, the township-level river chief and lake chief must be present at least once a week, and the village-level river chief and lake chief must be present at least twice a week. In each river patrol, the river chiefs of each river section shall carry out systematic inspections of ecological water treatment, water pollution prevention and control, bank protection and other management and protection measures of the relevant river section. River officers at the district level and above patrol the river to solve problems. Before conducting river patrols and surveys, river chiefs at county level and above may have the office of river chiefs at the same level and member units of the river chief system conduct overt and covert visits to grasp the outstanding problems in the rivers they inspect and understand the urgent and important problems that need to be coordinated and resolved by river chiefs at lower levels of river management. River leaders at the community and village levels are guided by the discovery of problems

Y. Hao and T. Wan, *The River Chief System and An Ecological Initiative for Public Participation in China*, https://doi.org/10.1007/978-981-99-4921-2_4

and focus on patrolling river sections where water management problems frequently occur, stopping actions that affect river water quality, and promptly reporting problems that cannot be solved by themselves. For river segments with many problems, the river superintendent in charge may code the frequency of patrols according to the needs of the job and keep a log of each patrol.

The field office is a common means for river chiefs of each river segment to discover, analyze, and resolve problems in the field resulting from patrols and investigations of the responsible river segment. The "Regulations for the Performance of the Duties of River and Lake Chiefs (Attempt)" issued by the Ministry of Water Resources emphasizes that provincial, municipal and county river and lake chiefs should be guided by problem solving and can work according to the actual situation in the field, coordinate and unify the opinions of all parties, consider problem solving measures and clarify problem solving requirements. There are two kinds of on-site office activities of river chiefs at all levels: One is to coordinate and solve the problems identified by river patrols on the river bank; the other is to hold an on-site office meeting of the river chiefs to clarify the most important and difficult problems in river management, promote on-site problem solving, and organize the next phase of key work. Participants in the on-site meeting of river supervisors usually include staff of the river supervisor's office at the same level, staff of relevant agencies at the same level and lower levels, and river supervisors at lower levels of the river section under control.

In practice, river chiefs at provincial, city, county and municipal levels independently decide on the research topic, the conduct of site visits and the form of site visits in accordance with relevant regulations issued by superiors and the same level for river chiefs visits. For example, on December 3, 2019, Hu Yacai, the river chief of Zhuganhe City in Xinyang City, Henan Province, presided over the on-site meeting of the river chief of Zhuganhe after inspecting wastewater treatment and river purification in Zhuganhe Basin, and put forward requirements on the implementation of the river chief system, water resources protection, water pollution control and other work tasks in Zhuganhe Basin. On December 17, 2021, Wu Hairui, the river chief of Xianyou County in Fujian Province, conducted an inspection and study of the systematic management of the Mulan basin in the county and made on-site arrangements for water pollution control work in the Mulan basin.

Under the hierarchical water management system, the river responsibility system builds a vertical chain of responsibility that passes the pressure layer by layer and condenses responsibility step by step, and a horizontal responsibility coordination mechanism that coordinates and unifies regions and departments. The vertical and horizontal responsibility relationship of local officials in water management refers to the power and responsibility relationship between the vertical and upper levels of government, and the horizontal responsibility relationship between regions and departments. With the introduction of the river basin management system, local governments can reorganize responsibilities through vertical contracting and horizontal coordination, and effectively adjust the responsibility system of local officials for water management. Reorganize rights and responsibilities by clarifying the relationship between rights and responsibilities, promoting vertical authority and promoting horizontal coordination, and clearly determining the relationship between the rights and responsibilities of local officials in water management; realize the reorganization of responsibilities through organizational improvement, improvement of rules and introduction of technology, and promote the efficient performance of responsibilities by local officials.

In order to realize the construction of ecological civilization, water resources management, as an important part of environmental management, must build and improve a modern water resources management system with Chinese characteristics. An important prerequisite for building a new water resources management system is to maintain the principle of scientific and efficient adjustment and redesign the responsibility system of local officials. The adjustment of vertical and horizontal relations follows the principle of block formation; the adjustment of vertical relations follows the principle of heterogeneity of tasks; the adjustment of horizontal relations follows the principle of division of labor and cooperation.

Under the water governnace system of "dual leadership", water conservancy includes water conservancy, environmental protection, housing, development and reform, land, health, agriculture, economy and trade, etc., which accept both the leadership of local officials and the vertical leadership of their own institutions. They should be accountable to both local chiefs and vertically led institutions. As a result, the daily efficiency of water treatment in the *"Tiao"* is low, the enthusiasm of the *"Kuai"* for water control is not high, and the phenomenon of evasion between *"Tiao"* and *"Kuai"* is common. In order to alleviate the contradiction between the Tiaos and improve the efficiency of water control, the ratio between *"Tiao"* and *"Kuai"* must be effectively adjusted.

"Tiao" is the product of the professional division of labor of governmental functions and focuses on solving problems in a specific area; *"Kuai"* as a level of governance comprehensively considers various problems within the area and emphasizes the systematic, holistic and creative nature of governance. In order to effectively coordinate the contradictions and differences between *"Tiao"* and *"Kuai"* in water governance and clarify the responsibility relationships, the river director system with the color of sports governance has transformed daily work into policy tasks, maintained the principle of *"Kuai"* as the mainstay, fully utilized the integration ability of *"Kuai"*, and used *"Kuai"* to lead *"Tiao"* in policy implementation. At each level of the *"Kuai"*, the overall river manager and the river manager responsible for a specific river basin are designated, the various departments are coordinated, and horizontal cooperation between departments is promoted.

Before the introduction of the river basin in charge system, local officials in the performance evaluation and GDP evaluation process tended to think about how to further increase GDP and improve policy performance, but were reluctant to pay too much attention to water management. When there is a problem with water management, local chiefs blame the vertical leadership of the water-related department and do not properly exercise the corresponding leadership responsibility. In order to fully mobilize the enthusiasm of local officials in water management, the river basin chiefs system builds a water management structure in which different levels of territorial governments are responsible for the principle of hierarchical responsibility and territorial management, and the responsibility for target management is given to Party and government leaders at all levels from top to bottom, who take the lead in the division of labor and cooperation among water management departments in the area. By strengthening assessment and accountability, local officials are encouraged to take the initiative and assume responsibility for water management in each administrative region. Under this pressure of assessment and accountability, local officials often strive to meet the goals and objectives of territorial water supervision in order to avoid being held accountable. In the case of Wuxi City, not only are river supervisors appointed at different levels, but also the results of water quality assessment are included in the performance evaluation of local party and government leaders, and if the water quality of the river in the jurisdiction deteriorates, the river supervisor is "vetoed" or even held accountable.

Under the isomorphic model of water management, the central government conducts vertical management, and government at all levels, from central to local, is very uniform in its functions, responsibilities, and institutional framework across all vertical levels. This not only makes it difficult to establish an effective hierarchical control system, but also leads to overlapping macro-control and micro-management functions between the central and local governments. As a result, responsibilities are not clearly divided between the central government and local governments at all levels of water management, leading not only to unclear boundaries of intergovernmental responsibilities, but also to local officials not being required to bear corresponding responsibilities in water protection even when they fail to fulfill their duties.

In order to clearly define the responsibilities of all levels of government in water management, it is necessary to move from isomorphism of duties to heterogeneity of responsibilities. The so-called heterogeneous responsibility means that, within the framework allowed by relevant national laws and regulations, local governments can introduce, through institutional reforms, a management system that is different from that of the central government or higher levels of government, depending on the natural conditions, economic characteristics, economic opportunities and other conditions on the ground. Based on the principle of heterogeneous responsibilities, the vertical water management functions of the government should be rationally adjusted, which mainly includes strictly distinguishing responsibilities between the central government and the local level, and between the higher and lower levels, so as to avoid overlapping functions; Local party and government leaders are the "main responsible" for water management in their jurisdictions, and directly carry out water management, rather than entrusting subordinates with water management; The beneficiary areas of water management and the level of government match one-to-one, and the level of affairs is precisely localized; Avoiding multiple levels of government for the management of the same water body and river segment, and introducing specific responsibilities for individuals to ensure that each river has a responsible river chief. Specifically, this means:

First, the responsibilities of supervisors and subordinates must be strictly separated and clarified by law. In order to effectively avoid chaos such as overlapping functions and conflicts between supervisors and subordinates caused by unclear authority and responsibilities at all levels, various municipalities have explored legislation and other methods to break the isomorphism of responsibilities and unify the responsibilities of supervisors and subordinates. For example, the Provisions on the System of River Chief System in Zhejiang Province, the first local regulation of the system of river chief system in China, takes a pioneering role by clarifying in legal form the differences in the division of responsibilities of river supervisors at all levels, with grassroots river chiefs at the city and village level focusing on conducting daily inspections of relevant waters and reporting identified problems, while river supervisors at the county or district level focus on encouraging relevant departments to resolve problems. Clearly defining the responsibilities of supervisors and subordinates in legal form is likely to overcome the disadvantages of the top-down and homogeneous responsibilities organizational forms in terms of vertical overlap of responsibilities, and helps to clarify the relationship of responsibilities between vertical levels.

Second, local governments focus mainly on directly addressing water pollution rather than leading subordinate agencies in controlling water pollution. In order to effectively avoid the governance risk of dispersed responsibility and the difficulties in pursuing responsibility under the collective responsibility of all levels, the opinions clearly propose to adhere to the linkage of party and government leadership and departments, and establish and improve the responsibility system with the responsibility system of party and government leadership as the core. In this governance framework, local party and government leaders at all levels no longer focus on managing subordinate water management, but take direct responsibility for water management and become the main actors of local governance networks. Local party and government leaders are integrating the forces of the various water supervision departments in their jurisdiction through guidance, coordination and supervision, changing the previous governance model of division of labor and responsibility at the department level, decentralized management to the model of chief responsibility, division of labor and cooperation at the department level (Cao & Zhou, 2019).

Third, the positioning of the transaction level should be accurate and appropriate so that the beneficiary area and the government level match. Combining river basin management and regional administrative management not only enhances attention to water issues at the local level, but also clarifies the responsibilities of the relevant party and government officials. In Jiangxi Province, for example, beneficiary areas, government levels and river basins have been combined, and a five-tier system of river managers at provincial, city, county, township and village levels has been established. At the provincial level, the secretary of the provincial party committee serves as the provincial river chief, the provincial governor serves as the deputy provincial river chief, and the respective provincial leaders serve as the provincial river chiefs of the major rivers and lakes to coordinate water management throughout the province. In the 11 county-level cities and the 100 counties and municipalities, the main leaders of Party committees and governments at all levels serve as chief river chief and deputy river masters in charge of water management in their administrative areas, including 88 city-level river masters, 822 county-level river chiefs, 2,422 township-level river chiefs and 13,916 village-level river chiefs. Through a relatively complete organizational system, the river chief system assigns the beneficiary areas of water management to the government level, which effectively promotes the enthusiasm and initiative of the people in charge of water management at all levels to perform their duties.

Fourth, it avoids the same water area and the same river section being managed at multiple levels of government. From the vertical point of view, the river basin chiefs system is a new kind of chief responsibility system, the core of which is the responsibility system, a government responsibility system based on party and government leaders at all levels, especially the main responsibility system of leadership. The river chief system refines the specific responsibilities for individuals through layers of responsibility and pressure transfer, and each river section has a corresponding river responsibility officer. For example, Zhejiang has taken the lead in introducing the river officer system throughout the province, and gradually introduced a five-level river chief system at the provincial, city, county, township and village levels, and formulated the mechanism of "one river, one leader, one team, one system and one handle to the end". On the one hand, it can highly stimulate the enthusiasm and initiative of leaders at all levels, adapt measures to local conditions and implement precise policies; on the other hand, it can unify systems, standards and technologies to facilitate daily assessment, monitoring and inspection.

Although the 2008 Water Pollution Prevention and Control Law clearly states that the State shall establish a responsibility system and an assessment and evaluation system for water pollution control objectives, and consider the fulfillment of water pollution control objectives as the content of the assessment and evaluation of local people's governments and their responsible persons, local officials should bear the responsibility for controlling water pollution within each administrative region. However, since river and administrative boundaries do not overlap, there are imperfect mechanisms of information flow and ecological balance between localities, as well as various obstacles in laws and regulations, management systems and technical procedures. It is difficult to hold local officials accountable for cross-regional water management issues. In addition, the fragmentation of functions between departments and the lack of effective division of labor and cooperation limit the effectiveness of local officials in water management. In order to effectively promote the performance of water supervision functions by local officials, it is necessary to pay attention to the division of labor and cooperation between departments and regions under the overall responsibility of local officials.

On the one hand, the division of labor and cooperation between departments must be maintained. Before the introduction of the river director system, local governments in each area were only nominally responsible, and the actual management was specifically responsible for the water-related departments of local governments at all levels, including environmental protection, water conservancy, housing, agriculture, forestry, development and reform, transportation, fisheries, marine and many other departments. These departments must accept not only the leadership of local governments but also the operational leadership of their higher-level counterparts, and there is no formal coordinating body, leading to difficulties such as separate management, difficulties in coordinating departments, and difficulties in cooperation. The model of river chiefs governing rivers has effectively implemented the transformation of government responsibility and division of labor between departments into the current leadership responsibility and departmental coordination, and the local party and government chiefs can achieve effective leadership and coordination as the axis of power exercise specifically.

As the chief responsible person of the local party and government, the river chief plays the coordinating role of the administrative authority through the river chief's office, the regional water resources management committee, the interdepartmental joint meeting system, etc., integrates the resources of relevant functional departments in water pollution control and resolves contradictions and conflicts between departments, overcomes communication difficulties, realizes institutionalized cooperation between multiple stakeholders across organizational boundaries, and effectively mitigates the conflict of interests between different functional departments of the government. At the same time, the first responsible person status gives the river manager the legitimacy to manage the river basin and can establish task indicators for the entire river basin. This institutional approach can effectively coordinate and integrate the departments involved in the management of the river basin's water resources, and carry out unified and coordinated management in accordance with the ecological natural law of the river basin's water resources (indivisibility of the river) to enhance the efficiency of management.

On the other hand, the division of labor and cooperation between areas should be noted. A watershed often flows through several administrative districts, and an administrative region usually includes several watersheds. Before the introduction of the river basin ladder system, local governments prioritized their own interests, and in the absence of mechanisms for coordinating interests and limiting opportunities, some local governments handled water management tasks passively, lacked enthusiasm for cooperative management, and found it difficult to cooperate across regions. The introduction of the river governance system has helped to solve the dilemma of interregional cooperation. At present, China's river basin management officer system is being implemented in provinces, municipalities, counties and other administrative regions, forming a mechanism for river management and protection at different administrative levels, providing an effective link between river basin management and county management, and enabling the coordinated development of the region's river basin economy and ecological environment. Under the river basin head system, provincial-level rivers that cross cities are headed by provincial heads, city-level rivers that cross counties (cities, districts) are headed by heads of county-level municipal governments through which the river basin flows, and county (district, city) level rivers that cross municipalities and city-government districts are usually managed by county (district, township) level river directors, respectively, and special village level managers, creating a four-tier river management and protection system consisting of provincial and township level river directors, county level river directors, township level river section managers, and village level specialists (neighborhood committee). At the same time, local governments have established organizational and institutional systems for trans-regional water management to better implement trans-regional water management. These systems effectively promote coordination, communication, dialog and cooperation among administrative regions, and provide an organizational and institutional foundation for strengthening cross-regional water management. For example, to further promote regional water control in river basins, the Zhejiang Provincial Water Resources Bureau has issued the "Guiding Opinions on Strengthening Inter-administrative Joint Water Control", which aims to establish and improve the inter-regional liaison and coordination mechanism of the "Five-Water Co-Governance", mutual review and mutual learning throughout the river basin, establish a mechanism for coordinated upstream and downstream water control with the borderland government as the main government in the province, and strive to form a pattern of coordinated and joint management of upstream and downstream and left and right bank.

In order to solve the phenomenon of unclear authority and responsibility and evasion between departments, and avoid the problem of inter-area cooperation in water management, local officials must take responsibility for horizontal coordination and solve the problem of local coordination and improve departmental coordination and unity by taking responsibility for local horizontal linkage and departmental horizontal coordination. The outstanding problem of comprehensive water management is that there are many water management departments in the same region that do not belong to each other. For the same river, water, soil, environmental protection, forestry, agriculture and other departments have different responsibilities, and it is not uncommon for departments to have unclear authority and responsibilities and to avoid each other. In order to effectively solve the dilemma of cooperation between departments, it is necessary to start from the following two aspects.

First, the responsibilities of departments must be clarified and the accountability relationships between departments must be clarified. With the introduction of the "River Chief System", a clear division of labor has been created in which each department performs its own tasks and responsibilities. For example, departments such as development and reform, environmental protection, water conservancy, agriculture, forestry and other departments should focus on the implementation and supervision of projects under their jurisdiction, and land departments should promptly study the land indicators required for the implementation of key projects and improve the situation that the environmental, economic and social benefits of the water basin cannot be developed synchronously due to the segmented management of departments. The "Work Plan for the Comprehensive Deepening of the River Guidance System in Yueqing City, Zhejiang Province (2017–2020)" clarifies the division of responsibilities of each municipality (road) and member unit. The provisions for strengthening the protection of water resources are as follows: the municipal water resources management bureau and the municipal development and reform bureau are responsible for implementing the most stringent water resources management system; The municipal water resources management bureau and the environmental protection bureau are responsible for conducting the pollution absorption capacity analysis of the city's water functional areas; The municipal water resources management bureau, the municipal comprehensive administrative law enforcement bureau, the municipal development and reform bureau, the municipal economic and information bureau, the municipal environmental protection bureau, the municipal public utility construction bureau and the municipal organ affairs management bureau are responsible for promoting the construction of a water-saving society; the municipal water resources management bureau and the municipal agriculture bureau are responsible for building rural water conservancy infrastructure.

It is important to improve interdepartmental coordination and build a unified coordination mechanism. Local Party and government leaders at all levels shall simultaneously act as river chief, take responsibility for organizing and directing the management and protection of relevant rivers and lakes, and build a unified coordination mechanism for water management through effective coordination, supervision, guidance, control and communication with the help of the comprehensive platform—the Office of the River Chief. In this way, it not only strengthens the relatively prominent position of environmental protection and water conservancy departments, but also clarifies the division of labor and responsibilities of functional departments such as development and reform, finance, public security, transportation, agriculture and forestry in water management, solves the problems of interdepartmental transfer, conflicts between policy goals and funds, lack of communication and omission of services, and promotes the fragmentation of sectional management. The example of Qinzhen in Shanghai shows that the local river management office, established according to the comprehensive coordination and promotion plan for river improvement in Qinzhen, is located half a level higher than other functional departments within the departmental level system, creating vertical pressure within the departmental level system to coordinate and mobilize the various functional departments. The head of the Office of the River Chief is the deputy mayor of Qin City in charge of river management, and the key members of the Office of the River Chief are the main officers of the various departments of Qin City. According to the documents of Qinzhen government, the main task of the river chief is to coordinate the river management tasks in Qinzhen, make an overall plan for river management, and coordinate various departments to jointly promote the management of river problems. In order to improve the coordination ability of the river director's office with other departments, Qinzhen has also adopted the completion of river management tasks and the work status of the river director's office as one of the end-of-year assessments of each department, establishing a rule basis for block-to-item coordination (Zhang, 2021).

The different locations of the headwaters and lower reaches of rivers and lakes, as well as the main tributaries, the different administrative regions to which the left and right banks belong, and the games of interest of different industries in the same place are important reasons for the poor management and protection of rivers and lakes. Therefore, the opinions specifically call for "clarifying the responsibilities for the management of rivers and lakes that cross administrative regions and carrying out joint prevention and control upstream and downstream and on the left and right banks" Establishing a regional liaison mechanism between local governments (provinces, cities, Counties and districts) in the "river head system" not only resolves the contradiction between the integrity of the river basin ecological environment and the fragmentation of administrative units, but also avoids the phenomenon of lack of cooperation and coordination between local governments in river basin management, and will also form a joint force for water pollution control.

Local governments at the provincial level fulfill the responsibility of horizontal linkage of transboundary rivers to avoid the problem of cooperation between municipalities in controlling water bodies. Joint prevention and control of rivers between provinces is the weak link in the current river governance system. In order to realize the coordination and linkage of inter-provincial river basins, the Central Leading Group for Comprehensive Deepening Reform took the Chishui River Basin as a pilot project to explore a governance model for joint work and cooperation in inter-provincial river basins. In February 2017, the Central Leading Group for Comprehensive Deepening Reform deliberated and approved the pilot plan for establishing environmental supervision and administrative law enforcement agencies by river basin, making the Chishui River Basin the only river basin where cross-provincial pilots are being carried out. The three provinces of Sichuan, Yunnan and Guizhou signed the "Chishui River Basin Environmental Protection Linkage Agreement", the "Yunnan-Guichuan Three Provinces Government Assistance in Promoting the Ecological Economic Development Cooperation Agreement of the Chishui River Basin" and the "Renhuai Declaration". The three provinces of Yunnan, Guichuan and Sichuan have successively carried out several joint law enforcement activities, uncovered and rectified more than 150 environmental problems, and jointly promoted the protection of the ecological environment in the Chishui River Basin. At the same time, various municipalities have also tried cooperation mechanisms such as the joint meeting system. For example, some local governments at the provincial level have jointly established and improved the leadership group and cooperation mechanism for Songliao water system protection, the information communication and coordination mechanism for water pollution emergencies in the Yellow River basin, the cooperation mechanism for water resources protection and water pollution prevention in Guiqian Province, and the joint conference system for water resources protection and water pollution prevention in the headwaters of the South-North Water Diversion Project (Hao, 2017).

Local governments at the provincial level should take responsibility for the horizontal connection of transboundary rivers to avoid the problem of cooperation between municipalities in controlling water bodies. Within provincial local governments, the existence of common higher levels of people's government enables the unification of local responsibilities, even in different local city and county governments. The People's Government of Zhejiang Province has actively explored the cooperation mechanism for inter-county water control and specifically issued the "Comprehensive Work Rules for Water Control in Zhejiang Province", which forms a system and standards for inter-regional water control and ensures that there are rules for water control. Articles 18 and 9 are as follows:

Article 18: The people's governments of the county-level cities and counties (cities and counties) bordering the same river basin shall establish consultation and coordination mechanisms for water control, rationally develop and utilize water resources, and jointly carry out the following work:

(1) *Strengthen information exchange, immediate exchange and sharing of relevant information, such as water monitoring work agreements, assessment reports and emergency warnings;*
(2) *Negotiate and dock treatment measures to prevent and control water pollution damage and other water hazards in all administrative regions;*
(3) *Carry out water rights trading and other cooperative activities.*

When the people's governments of county-level cities, counties (cities and counties) and their relevant departments find that a large area of flotsam has appeared on the surface of the river, they shall immediately organize interception and clean-up, investigate the source and notify the relevant people's governments and their relevant departments, such as the places of inflow and outflow; the relevant people's governments and their relevant departments shall immediately organize personnel, take measures and jointly participate in treatment.

Article 19: In the efforts for water regulation in river basins covering several administrative regions, the people's governments at a higher level in the region in which they are located shall establish a joint mechanism for coordinating prevention and control and coordinate planning, functional zoning, major projects and construction and operation of monitoring and control facilities related to water regulation in the same river basin within the region.

If the relevant people's governments cannot reach consensus on water regulation measures in administrative regional river basins through consultations, their joint higher-level people's governments shall coordinate and adopt such measures without delay.

In order to improve the effectiveness of local chiefs in water management, the river chiefs' system needs to transform the power and responsibility relationship of local chiefs in water management. In this regard, the river chiefs system clarifies unclear public responsibilities by clarifying powers and responsibilities; through authoritative promotion, the vertical implementation of local officials' responsibility in water management is realized; through cooperative management, the horizontal responsibility and cooperation of local officials in water management can be achieved. The river responsibility system precisely specifies the responsibilities of different subjects for water management, determines the requirements for the management of different rivers in detail, and precisely assigns the specific tasks of each responsible person by dividing responsibilities, filing and making a list of responsibilities to clarify the unclear responsibilities.

First, the responsibilities of each level are subdivided. The proper operation of the river management system depends on the hierarchical and hierarchical operation of the bureaucratic system, that is, according to the principle of territorial division and hierarchical management, the responsibility for water control, which is an integral public matter, is divided and tailored, and each administrative level and administrative area is assigned a specific responsible person, creating a hierarchical operation with clear responsibilities at each level. The responsible bodies at all levels have different responsibilities and duties, with the provincial river chief generally in charge of directing water control work in the province and taking responsibility for general supervision and overall planning. River chiefs at city, county and township levels are responsible for organizing and directing specific management and protection activities, coordinating and solving relevant problems, supervising the performance of tasks by relevant departments at the same level and responsible agencies at the next level, and evaluating the achievement of targets. River officers at the village level are mainly responsible for publicity, inspection and control of river and lake protection. The opinions specify the responsibilities of river and lake officers at all levels as follows:

(1) *River chiefs of the provinces and chiefs of the lakes. The river chiefs of each province (autonomous region or municipality directly under the central government) shall be responsible for the management and protection of the rivers and lakes in their respective administrative areas and shall make general arrangements, coordinate, supervise and evaluate the management and protection of the rivers and lakes in their administrative areas. Provincial river and lake chiefs are mainly responsible for organizing and carrying out special measures to rectify conspicuous problems on rivers and lakes, coordinating and solving major problems in the management and protection of rivers and lakes in charge, reviewing and organizing the implementation of the "One River (Lake)—One Policy" plan for rivers and lakes in charge, coordinating and clarifying the responsibilities for the management and protection of rivers and lakes in the various administrative regions, promoting the establishment of inter-regional and interdepartmental coordination and linkage mechanisms, and monitoring and evaluating the performance of annual tasks by provincial competent departments and river and lake managers at the next level.*
(2) *River and lake chiefs at the city and county level. The river chiefs of cities (prefectures) and counties (districts) shall be responsible for the management and protection of rivers and lakes in their respective administrative areas. The city chiefs and county chiefs are mainly responsible for implementing the higher-level operation of the chiefs of rivers and lakes; they conduct daily inspections of the rivers and lakes in charge and organize the timely rectification of problems; Review and organize the implementation of the "One River (Lake)—One Policy" plan for the responsible rivers and lakes, and organize and carry out special treatment work and special corrective measures for the responsible rivers and lakes; coordinate and stimulate the responsible departments to formulate and implement plans for the management, protection and administration of the responsible rivers and lakes, and coordinate and solve major problems in the implementation of the plans; Supervise the formulation of a list of responsibilities of departments belonging to the system of river and lake chiefs at the same level, and promote the establishment of inter-regional coordination and linkage mechanisms; supervise and request river and lake chiefs*

at the next level and relevant departments at the same level to deal with and solve problems arising on the responsible rivers and lakes, investigate relevant illegal acts, and deal with them in accordance with the law, and supervise and evaluate their performance of duties and completion of annual tasks.

(3) *River and lake chiefs at the municipal level. The river and lake chiefs at the municipal level shall be primarily responsible for carrying out the work assigned by the river and lake chiefs at the higher level and implementing the specific tasks of river and lake management and protection; conducting daily inspections of the responsible rivers and lakes and arranging for the rectification of problems found during the inspections; and promptly reporting to the river and lake chief at the next higher level or the relevant departments through the river and lake chief at the next higher level. The river and lake chiefs at the village level, appointed in accordance with local conditions, shall be mainly responsible for publicizing the protection of rivers and lakes among villagers, organizing the conclusion of village rules and regulations for the protection of rivers and lakes, and conducting daily inspections of relevant rivers and lakes, conducting daily inspections of relevant rivers and lakes, discourage and stop violations of laws and regulations in rivers and lakes, promptly resolve such violations that can be resolved and report them to the relevant river and lake chiefs or departments at the next higher level if they cannot be resolved, and cooperate with relevant departments in field law enforcement and investigation and handling of disputes related to rivers and lakes (investigation support), etc.*

The second option is to divide the responsibilities of each department. In China, the responsibilities of water environment administration are divided among local departments of water conservancy, environmental protection, housing, land resources and other departments, and the responsibilities and cooperation between departments are not clear, resulting in the accumulation of water resources, water ecology, water environment and other problems. Without breaking the current institutional structure of the "Nine Dragons Managing Water," the river officer system shifts management responsibility upward and takes full advantage of the authority and resources of local party and government leaders to coordinate and integrate the dispersed forces of departments. The river basin officer system has shifted responsibility for overall management upward, while specific management work continues to be shifted downward, that is, river and lake management continues to depend on the implementation of the functions of various departments. The transformation of the river management system from "Jiulong water management" to a management system with overall management and division of labor and coordination among departments is the transformation of water environment management from "decentralized management by departments" to "responsible management and division of labor and coordination among departments". Implement the corresponding legal system everywhere. Classify and guide environmental protection responsibilities and clarify relationships. For example, Zhongshan County in Liupanshui City, Guizhou Province, has refined and quantified the overall goals and tasks of rehabilitation, and divided in detail the responsibilities of county supervision, finance, auditing, environmental protection, development and reform, economic and information technology, housing, municipal government, agriculture, water management, industry and commerce, energy supply, forestry, health and food and drug administration, and cities and towns. The People's Government of Zhongshan County has signed a target responsibility statement with all the above-mentioned departments, cities, townships, subdistricts and large and medium-sized state-owned enterprises located in the county, detailing the target tasks and work requirements (Ren, 2015).

2 One River One Policy

In December 2016, the General Office of the CPC Central Committee and the General Office of the State Council issued the Opinions stating, "Based on the actual conditions of different regions and different rivers and lakes, coordinate the upstream and downstream, left and right banks, implement a policy for one river and a policy for each lake, and solve outstanding problems in the management and protection of rivers and lakes. In September 2017, the Ministry of Water Resources issued the Preparation Guide for the "One River (One Lake) One Policy" Program to standardize and guide the implementation of the One River One Policy, One Lake One Policy" work and support the implementation of the river management system. In December 2017, the General Office of the CPC Central Committee and the General Office of the State Council issued the Guiding Opinions on the Implementation of the Lake Chiefs System (hereinafter referred to as the Guiding Opinions), which set out clear requirements such as "the establishment of a river (lake) and a file/policy".

The preparation of one river (lake) policy plan is based on the general task requirements of "Guiding Opinions", taking into account the real and important problems and causes of rivers and lakes at all levels, sets the goals of management, protection and management, proposes measures for management and protection, and provides action plans for the management, protection and control of rivers, lakes and river sections at all levels for the work related to the river governance system. On the one hand, the formulation of a plan for a river (lake) policy is conducive to the clarification of the specific tasks of each river and lake administration. In formulating the plan for a river (lake) and policy, it is necessary to follow the principle of problem-oriented, overarching coordination, step-by-step implementation and clear accountability that is stipulated in the preparation guide, and on the basis of the preliminary investigation, based on the characteristics and current situation of the river and lake, clarify the management objectives, management tasks, management priorities, etc., and establish a feasible schedule for the countermeasures for improvement. Gradually implement rivers with heavy pollution and major treatment difficulties; focus on comprehensive elimination of pollution sources such as river discharges for rivers in watersheds and regional main rivers; focus on river dredging and construction of supporting water conservancy facilities for major rivers in counties and municipalities; focus on river dredging and ecological treatment and rehabilitation of rivers in cities and villages. For example, Beijing Municipality has instructed the Academy of

Water Sciences to formulate the "Guidelines for the Preparation of the "One River, One Policy" Plan for the Beijing River Guidance System (Trial)" for 14 river basins, including the Canal Basin and the Yongding River Basin, and to instruct all counties to prepare the 2018 Annual Plan and the 2018–2020 Three-Year Plan at the county level. The 14 river basins are different, and the focus of management is very different. Miyun Reservoir, Guanting Reservoir, Chaobai River Basin, etc. are mainly for water source protection; North Canal, Qinghe River and Liangshui River Basin are mainly for wastewater treatment; Yongding River Basin focuses on ecological restoration and protection.

The preparation of a river (lake) policy plan serves to implement specific management tasks, which are assigned to the responsible units and persons in turn. The content of the river (lake) policy plan includes the basic situation and existing problems of rivers and lakes, the requirements for management and protection objectives, tasks and indicators, the implementation of specific objectives for rivers and lakes (river sections) and tributaries, as well as the countermeasures and planning arrangements for management, management and protection. This is depicted as "Five and Two" program, consisting of five lists such as problem list, target list, task list, action list and responsibility list, and two targets such as targeting decomposition and strategic arrangement target. The "One River (One Lake) One Policy" plan not only systematically sorts out the main problems and causes of rivers and lakes, sets out the management and protection goals and main tasks for rivers and lakes, but also proposes targeted and feasible measures and the corresponding responsible persons and responsible agencies. According to the requirements of hierarchical management of the system of river and lake managers, the overall objectives, main tasks and control indicators of river and lake management and protection are divided among each section of each river, and the specific responsibilities of each river manager at each level and section are clarified in terms of publicity, delimitation, obstacle removal, cleaning, pollution prevention and treatment. Based on the targets and implementation schedule, the responsibility for management and management is gradually transferred to the responsible units and individuals, so that the content of the responsibility for river basin management moves from general plan to specific policies and implementation files.

The establishment of a river (lake) and implementation files is helpful in accurately determining the water management responsibility of each river. Since the specific conditions of each river and lake are different, the main problems and the difficulty of solving them are also different, which requires a specific analysis of the specific problems rather than a uniform approach to management. The introduction of the action plan "one river and one policy/file" firstly means that the most basic condition survey and data collection must be carried out at the city, county and township level, including the basic situation of the river, the basic situation of the lake, the water quality of the river and the lake over the years, the type of the main functional area of the river and the lake, and for the rivers and lakes belonging to the water functional area, the evaluation of the most strict water resource management system should also be registered. Although red tape and bureaucracy has a negative effect on organizational effectiveness, organizational developmental objective alleviates the negative effects of bureaucratic red tape (Pandey et al., 2007).

Based on surveys at city, county and township levels, the basic situation of all rivers and lakes in the province will be connected into an information network, a network information database will be established, and monitoring networks and monitoring maps will be made. For some rivers crossing administrative regions, overall planning is carried out and multiple parties are considered. "One river (lake) one file" creates a file for each specific object to identify and master the basic number and current situation of river and lake systems, clarify the responsibility for managing rivers and lakes, and provide important basic information for the condensation of river and lake managers at all levels to carry out their duties.

A river (lake) and a file general ledger information consists of two parts: basic information and dynamic information of rivers and lakes. Basic information includes the information on the natural characteristics of rivers and lakes and the information on the length of rivers and lakes. The dynamic information mainly reflects the water volume, water environment and aquatic ecology of rivers and lakes, and the development, use, management and protection of rivers and lakes. One river (one lake) and one file, on the one hand, provide an information base for the division of labor in water management. The "one river (lake) one policy" system captures the current situation and dynamic changes in river and lake management and protection by creating dynamic information books for river and lake water resources, shoreline, water environment, and aquatic ecology. It helps solve practical problems, such as the lack of clarity of the river system and hierarchy, the lack of understanding of the basic situation, functional positioning and key issues of rivers and lakes/river sections, and the lack of clarity in the information and responsibility requirements of those responsible for river and lake management and protection, and provides comprehensive information support and technical basis for the comprehensive implementation of the river (lake) management system and the effective division of labor in water protection tasks. Thus, Fangxian Water Bureau organized the county's technical staff to conduct a comprehensive survey of the county rivers, register and compile records, and create archives in various forms. Based on the river maps issued by the state and higher authorities, it was verified that the information on the rivers in Fangxian County's jurisdiction is in line with the actual situation, and the office of the head of the Fangxian River Working Group was established. At the same time, according to the actual needs, the staffing is determined, the work tasks and responsibilities are specified, and an orderly division of labor is made.

3 Formal and Informal Water Governance Institutions

Daily monitoring by the public is the critical link to achieving river long-term governance. As part of the comprehensive implementation of the river head system across the country, the public can provide bottom-up feedback on water control issues through multiple monitoring channels of formal and informal institutionalization and influence higher levels of government power to promote effective monitoring. As stakeholders in the formulation, implementation, and monitoring of public policies, the public can not only be the first to perceive the urgency of introducing policies and the effectiveness of implementation, but also identify deviations in the implementation of policies in a timely manner. The management of rivers and lakes is closely related to the lives of citizens and is a social problem that citizens can personally perceive. In the process of realizing the rule of the river chiefs, the public is an important subject in monitoring the river master's management. Based on the identity characteristic, the public can be divided into two types: ordinary citizens and civil river chiefs. Ordinary citizens can use their geographical advantage of living and working near rivers to be the first to grasp and collect the dynamic situation of rivers and water bodies, which compensates for the low frequency of inspections by government river supervisors and the insufficient collection of information on rivers. Civilian river chiefs such as village river chiefs hired by the government can use their own knowledge and professional advantages to actively collect samples and test water quality, report the results to the state river supervisor in time, and be the proper assistant to the state river supervisor.

As for the objectives of supervision, at the beginning of the introduction of the river supervisor system, all provinces, autonomous regions and municipalities directly under the central government have established an organizational system of river supervisors at the provincial, municipal, county and city levels in accordance with the requirements of the opinions. In 2017, in order to develop the "last mile" of river and lake management, Guizhou Province took the lead in extending the organization system to villages based on the four levels of river chiefs at the provincial, municipal, county and city levels, and achieved full coverage of the five-level river chief system, which not only achieved remarkable results in the treatment of black and odorous water, but also quickly solved the problem of river and lake impoundment. Subsequently, more than 20 provinces, autonomous regions and municipalities directly under the central government followed suit and introduced the five-tier system of river chiefs at the provincial, municipal, county, city and village levels. Among the five-tier river chiefs, the village-level river chiefs are different from the state-level river chiefs at the municipal level and above: On the one hand, due to the pressure-based system, the village-level river chiefs undertake the task of issuing contracts and demarcating them from the higher-level river chiefs; on the other hand, due to the pressure from the electorate, the village-level river chiefs have to respond directly and immediately to people's demands, while the village-level river chiefs have other daily tasks to perform. This eventually led to an overload of village-level river chiefs who were unable and unwilling to perform their duties. Legally, administrative villages are not first-level administrative organs, village committees are run by municipal governments as autonomous organizations, village party secretaries and village committee directors are not civil servants, they are not covered by the Civil Service Law, there is no administrative accountability, and municipal governments (neighborhood offices) can only sign environmental management agreements with them, which greatly weakens the effectiveness of the river chief system's accountability. The difference between village-level river chiefs and government river chiefs' results in the need to open multiple channels to ensure effective public scrutiny of the five-tier river chiefs.

In order to protect their legitimate rights and interests, people have the right to make legitimate demands on the agent—the government—and to actively provide relevant information to the government through various channels so that they can remain alert to problems in public affairs. The government takes the initiative to develop and establish a variety of institutionalized channels, encourages the public to participate in water control, and plays the role of the public in monitoring. At the same time, in addition to formal institutionalized monitoring channels, the public also assumes informal accountability to monitor and hold river chiefs accountable.

The number and unimpeded availability of formal institutionalized monitoring channels is the basic prerequisite for the public to effectively monitor river chiefs at the community level and beyond. As part of the implementation of the river chief system, local governments at all levels have erected public signs on riverbanks (rivers and river sections) in accordance with the requirements of the opinions, providing information on the river channel, the name of the river steward, his affiliation, his contact number, the content of management and protection, and so on. In addition, the "Implementation Plan of the Ministry of Water Resources and the Ministry of Environmental Protection for the Implementation of the "Opinions on the Comprehensive Implementation of the River Chief System" proposes to make extensive use of various media and communication tools such as newspapers and magazines, radio, television, Internet, WeChat, Weibo and customers... Widely publicize and guide, and continuously strengthen the public's sense of responsibility and participation in the protection of rivers and lakes. In practice, considering local conditions, local governments have used advanced network technologies to explore and set up river chief information platforms, WeChat platforms, and other types of APP to provide convenient and diverse channels for public supervision. For example, in order to strengthen the public's role in monitoring and accountability in the implementation of the river chief system, Guangdong Province has established public complaint and reporting platforms, such as the public WeChat account "Guangdong Smart River Chief", the WeChat mini-program, the river chief's 24-h hotline, and the river chief's mailbox "Guangdong Provincial Affairs". According to statistics, the "Guangdong Smart River Chief" platform received a total of 1,301 complaints in 2021, with a completion rate of 97.9%.

Tsai's comparative study in four villages in Jiangxi and Xiamen found that informal accountability exists in Chinese rural society, which can effectively compensate for the lack of formal accountability. The so-called informal accountability refers to the community organizations formed within the village based on common interests and moral obligations-usually with common blood, beliefs, or interests, and there is a relationship of moral responsibility and obligation among them-can exercise moral accountability to villagers (especially village leaders) and encourage villagers to provide public goods for community organizations. If villagers do not fulfill their collective accountability, they lose their moral status, are excluded from group activities, or even boycotted by community organizations, making it difficult for them to continue their work in the village. Currently, maintaining moral status is still very important for village leaders who direct and carry out work in the village. Moral accountability plays an important role in the behavior of village leaders (Tsai, 2007). The same is true for the implementation of the river chief system at the village level. On the one hand, given the common demands of villagers for water control, village heads who were born and raised in Sri Lanka will perform their duties conscientiously under the pressure of moral accountability to gain the recognition of villagers, even in the absence of administrative accountability from higher levels of government. On the other hand, in addition to water control, village leaders must also assist the local government in performing other tasks that are inseparable from assisting villagers. In order to reduce the difficulties in carrying out the work in the future, village leaders will also do their best to meet the villagers' demands for water control.

With the help of diversified monitoring channels, citizen complaint has increasingly become the most widely used monitoring method of individual citizens. To ensure the effectiveness of citizen complaint monitoring, various municipalities have gradually established a system for handling complaints and reports under the river master system. For example, in 2018, the Anhui Provincial Water Resources Department issued the "Opinion on Strengthening the Comprehensive Implementation of the River Chiefs' Complaint and Report System", which calls on river chiefs' offices at all levels and member units of the River Chiefs' Assembly in the province to designate special personnel to handle phone calls, letters, emails, faxes and receive visits for complaints and reports, and set clear provisions on the handling process, completion time, reward and punishment measures, etc. In order to mobilize the public's enthusiasm to participate in monitoring, some provinces and cities have also established a mechanism to reward reports, such as Tianjin City's report to the public, which is rewarded with a one-time reward of 200 yuan after validity is verified.

However, citizen complaint monitoring is essentially an interactive form of participation, and the key to monitoring effectiveness lies in the government's timely and effective response to problems complained about by the public. In China's administrative bureaucracy, the power of leading cadres increases with the administrative level at which they are located, and the higher the administrative level of officials, the more power they have, the more public resources they can mobilize to solve social problems, and the more effective and credible their responses. The hierarchy of the power system that can be achieved through citizen political participation is directly related to the degree to which the purpose of political participation is achieved. In the river head system, local party and government leaders who serve as river heads represent the higher level of power in local government. In monitoring river management and protection, the public identifies and reports problems to the river chief, and if the river chief at the same level cannot solve them, they report them to the river chief at the higher level, who coordinates and solves them. In this way, the public can directly or indirectly influence the higher level of power of local governments, drive the resolution of water control problems, achieve public monitoring in the true sense, and improve the effectiveness of monitoring.

References

Cao, X., & Zhou, J. (2019). River chiefs promote good governance in river basins: What is possible and what can be done. *Jiang Hai Academic Journal* (in Chinese) (6), 139–148.

General Office of the Ministry of Water Resources. (2019). *Opinions on further strengthening the performance of duties and responsibilities by river and lake chiefs*. Beijing Ministry of Water Resources. https://www.gov.cn/zhengce/zhengceku/2020-01/01/content_5465754.htm

Hao, Y. (2017). The responsibility of local officials in the river chief system. *Journal of Henan Normal University (Philosophy and Social Sciences Edition)* (in Chinese), *44*(5), 13–18.

Pandey, S. K., Coursey, D. H., & Moynihan, D. P. (2007). Organizational effectiveness and bureaucratic red tape: A multimethod study. *Public Performance & Management Review, 30*(3), 398–425. https://doi.org/10.2753/PMR1530-9576300305

Ren, M. (2015). "River Chief System": A sample study of cross-departmental coordination in river basin governance of the Chinese government. *Journal of Beijing Institute of Administration, Journal of Beijing Institute of Administration* (in Chinese) (3), 25–31.

Tsai, L. L. (2007). Solidary groups, informal accountability, and local public goods provision in rural China. *The American Political Science Review, 101*(2), 355–372.

Zhang, G. (2021). Using the section layer against the bureau layer: The operating logic, internal tension and institutional resilience of the river chief system. *Journal of Tianjin Institute of Administration* (in Chinese) (1), 55–64.

Chapter 5
Dual-Governance Structure and Forms of Responsibility

1 When Top-Down Meets Bottom-UP: Vertical and Horizontal Governance

Through the system of vertical responsibility of river chiefs, the river chiefs' system promotes the active efforts of local chiefs to implement water management objectives, and forms a pattern of water management based on "attention of "leadership, joint grasping and joint management". These remarkable achievements in water management depend on the authoritative forces of vertical promotion, with high-level design, high-level promotion and leadership as important ways to promote the vertical promotion process.

Due to the unclear positioning of the functions of the Chinese government, problems often arise in water conservancy, such as overlapping responsibilities between the central government and local governments, conflicting responsibilities between "Tiao" and "Kuai", unclear jurisdictional boundaries between regions, and functional fragmentation between departments. In order to clarify the relationship between the rights and responsibilities of local officials in water management and improve the efficiency of water management, the River Chief System is comprehensively rebuilding the water management system from the top planning level to solve the symptoms and causes of water management.

To protect and improve the environment, prevent and control water pollution, protect aquatic ecology, ensure the safety of drinking water, maintain public health and promote the construction of ecological civilization. On June 27, 2017, the 28th Session of the Standing Committee of the Twelfth National People's Congress amended and adopted the Law of the People's Republic of China on the Prevention and Control of Water Pollution. In the form of standardization and legalization, it clearly requires that local people's governments at all levels shall be responsible for the quality of the water environment in their respective administrative areas and take timely measures to prevent and control water pollution, and points out, that provinces, cities, counties and municipalities shall establish a system of river commissioners and organize and direct the work of water resources protection, riparian management,

Y. Hao and T. Wan, *The River Chief System and An Ecological Initiative for Public Participation in China*, https://doi.org/10.1007/978-981-99-4921-2_5

water pollution prevention and control, and water environment management of rivers and lakes in their administrative areas. The revision of this law by the state fundamentally contributes to solving problems such as unclear authority and responsibility between superiors and subordinates, division of regulations and evasion between departments at the same level. This not only clarifies at the level of national law that the main responsibility for water management lies with the main party and government leaders at all levels at the local level, but also establishes legal powers for local party and government leaders at all levels to effectively coordinate and manage the various departments of water management, and reshapes the system of water management. At the same time, the local ordinances also clearly formulate the requirements for the implementation of the river basin management system, which provides further legal guidance for the reconstruction of the water management system. The local ordinance, which was passed by the 12th Session of the Standing Committee of the 12th People's Congress of Zhejiang Province, is the first special legislation on the provincial-level river basin management system in China, clearly setting out the working organs of the river basin management system at or above the county level and the working responsibilities of river basin managers at all levels, and providing a legal basis for clarifying the responsibility relationships between levels and departments in the water management system and reconstructing the water management system.

The river chiefs model reaffirms the relationship between the rights and responsibilities of party and government leaders in water management. The opinions clearly call for the establishment of a four-tier system of "provinces, cities, counties and municipalities," and that each province (autonomous region or municipality directly under the central government) should establish a general river chief, with the comrade in charge of the Party committee or government as the comrade in chief; In each province (autonomous region or municipality directly under the central government), river officers shall be established on the main rivers and lakes in the administrative areas, with the comrades-in-charge at the provincial level acting as comrades-in-charge; in the cities, counties and municipalities where rivers and lakes are located, river officers shall be established at different levels and in different sections, with comrades-in-charge at the same level acting as river officers. River chiefs at the county level and above shall establish corresponding river chiefs' offices, and the specific composition shall be determined by each locality according to actual conditions. In practice, various localities continue to innovate and gradually develop the organizational structure of the five-level linkage province-city-county-municipality (town)-village, creating the integration effect of the five-level linkage. At present, the key chairmen of Party committees and governments of 31 provinces (autonomous regions and municipalities directly under the central government) serve as provincial river commissioners, a total of 300,000 provincial, city, county and municipal river and lake commissioners, and more than 900.000 river and lake commissioners at the village level (including river patrols and river guards), so that river and lake management and protection responsibilities are fully covered, and an organizational structure has been established in which local Party and government leaders at all levels are responsible as a whole.

According to the general requirements of the opinions, the river head system is systematically rebuilding the accountability system by reshaping the accountability relationships and transforming the organizational structure. Under the influence of the vertical accountability system, various localities have created a top-down, layered and interlocking administrative accountability system for river chiefs. On the one hand, the main tasks should be refined and implemented, the classification guidelines should be strengthened, the work progress should be clarified, and the responsibilities of river supervisors at all levels should be refined, quantified and specified; on the other hand, the relationship between the rights and responsibilities of supervisors and subordinates is clearly divided, so that the power and responsibility and financial power and responsibility correspond to the responsibility, and a complete and clear vertical relationship is established. In this way, those in charge of rivers and lakes are encouraged to actively implement the six major tasks of strengthening water resources protection, riparian management and protection, water pollution prevention and control, water environment management, ecological restoration, law enforcement and supervision, and establish a chain of responsibilities for river and lake management and protection that runs from top to bottom.

The comprehensive implementation of the river management system is an inherent requirement for promoting the construction of ecological civilization, and it is also a decision and commitment carried out by the state in conjunction with the national conditions and water conditions, and is highly appreciated by the central and local governments. Under the active encouragement and promotion of the central government, the key leaders of local parties and governments at all levels serve as river chiefs in their administrative regions, which can not only effectively gather political authority and social resources in the area under their jurisdiction, but also gradually pass on responsibility pressure in the bureaucratic system, form a "commanding water leadership" promoted by a high position, release enormous political potential energy, and have a huge promotion and incentive effect on the implementation of the river chief system.

On the one hand, the central government promotes the implementation of the river basin management system at the local level. Since the 18th National Congress of the Communist Party of China, the central government has promoted the construction of ecological civilization and the management and protection of rivers and lakes with unprecedented intensity. In November 2016, the central government decided to fully implement the river head system throughout the country, and the river head system was elevated from local practice to national will. In his 2017 New Year's message, General Secretary Xi Jinping emphasized that "every river must have a river chief" and issued the strictest order to fully implement the river chief system. On June 27, 2017, the 28th Session of the Standing Committee of the Twelfth National People's Congress amended and adopted the Law of the People's Republic of China on Water Pollution Prevention and Control, which clearly stipulates the establishment of a river stewardship system at the provincial, municipal, county and city levels in a standardized and legal form. The Ministry of Water Resources has also established a leadership group to promote the river officer system, which is headed by the minister and is responsible for strengthening the organization and leadership of the work to promote the river officer system, guiding and urging all municipalities to fully implement the river officer system, and coordinating and solving the major problems in the implementation of the river officer system. In addition, the Ministry of Water Resources has established a supervision and inspection system to guide and urge all municipalities to strengthen organizational leadership, improve work mechanisms and implement work responsibilities; organize training courses on the river officer system to improve the policy level and operational ability of grassroots river and lake area managers and the backbone of river area managers; The "Regulations for Fulfilling the Duties of River and Lake Chiefs (Attempt)" was issued, detailing the duties and tasks of river and lake chiefs at all levels and other measures to improve the ability of river and lake chiefs to fulfill their duties and accelerate the promotion of the river and lake chief system. By continuously transmitting the political signals from the top level of the central government downward, the attention of party and government leaders at all levels has been drawn to the river and lake chiefs system, a strong "political potential" has been created and transmitted to the grass-roots government through the vertical hierarchical structure, greatly promoting the development and implementation of the river and lake chiefs system in various places.

A particular political hierarchy, political signal, or political representation may motivate local officials to actively implement public policies. When the system of river managers was established, the local party and government leaders, as the chief river managers, coordinated the various functional departments to pool administrative powers and resources, and applied the method of high-level promotion and layer-by-layer decomposition to transfer the task of river management to the local authorities at a lower level. In the bureaucratic system, local governments mainly ensure the implementation of the "responsibility system" of local party and government leaders by quantifying the downward pressure of governance indicators and integrating institutional resources. First, the quantitative downward pressure of indicators. Higher-level governments formulate river and lake management targets according to the overall development of the region. To ensure that the goals and objectives are met and to monitor management at the grassroots level, they will often quantify them in terms of specific indicators and pass them on to grassroots management. The layer-by-layer quantitative breakdown of indicators improves the relevance of grassroots officials' actions, and the pressure of indicators forces them to act to fulfil higher-level tasks, and the fulfilment of higher-level index tasks also has a positive incentive effect on lower-level officials, which can make them actively promote the work of the river head system to achieve incentives. The second point is the integration of institutional resources. In promoting the implementation of the river head system, the grassroots government relies on the political potential energy formed by the top-level political authority and the attention of the higher-level government, and obtains more financial resources to build the river head system and invests them in the management of rivers and lakes. In 2018, for example, County C actively sought the support of higher authorities and relied on resources within the system, and launched 42 river and lake pollution control projects within the county's financial plan and invested 1.022 billion yuan; at the same time, with the support of all parties, 191 new projects were added outside the financial plan with an additional investment of 810 million yuan. From the point of view of resource utilization, the strength of political potential greatly affects the amount of resources obtained within the grassroots government system, and also influences the ability and enthusiasm of subordinates to implement the responsibility system.

For a long time, local governments in the river basin were independent and lacked common prevention and control; water environment management functions are scattered, with unclear responsibilities and poor coordination between departments; supervisors and subordinates are "isomorphic in their responsibilities" and the boundaries of responsibility are blurred. This requires governments at all levels to coordinate and integrate previously dispersed administrative forces, effectively solve the fragmentation problem in water management, and manage holistically. Under the decentralized management system of division of labor in China, without changing the current territorial management system of hierarchical responsibility and division of labor among departments, local party and government leaders serve as river heads and make full use of their own powers and resources to coordinate and integrate the various government forces.

The core idea and foundation of the river chief system is that Party and government leaders serve as river chiefs, and this system is in line with China's political tradition of "high-level leadership through command and coordination." As a political tradition, the mechanism of leadership is to make full use of the authority and resources of the leadership to coordinate and integrate the decentralized government forces. The first task is to unify the interdepartmental administrative forces. Under the river commissioner system, local party and government chiefs serve as river commissioners and act as the first responsible person to organize, coordinate, guide and supervise the management of the relevant water areas, realizing the transformation of water environment management from "decentralized management by departments" to "responsible management by chiefs and departments with division of labor and cooperation." For example, at the provincial level in Jiangsu Province, not only the River Responsible Bureau has been established, but also a River Responsible System Leadership Group has been set up, which strengthens interdepartmental communication and jointly explores and coordinates solutions to important and difficult problems in water environment management through regular meetings of the River Responsible Working Bureau and liaison meetings of the members of the Leadership Group.

River chiefs monitor the accomplishment of the tasks of the relevant departments and river chiefs at the next lower level, evaluate the achievement of goals and objectives, and strengthen incentives and accountability. In accordance with China's political system, the river chief system divides responsibility for water control from top to bottom and establishes the corresponding level of responsible persons according to the administrative level of the river basin fragment, thus creating a "chain of responsibility" that transmits pressure and is accountable at each level. Through the supervision of the river manager at the lower level, the river manager at the higher level pushes the river manager at the lower level to actively carry out its responsibility and fulfill the task of water control. For example, the Office of the River Commissioner of Wuxi City regularly reviews and reports on the implementation of the river commissioner system in each county and district, and publishes the evaluation results and rankings; Foshan City in Guangdong Province surveyed a number of key river commissioners at the county and district levels on the pollution cleanup of the Guangzhou-Foshan transboundary rivers; in 2017, Jiangmen City reported and surveyed more than a dozen river commissioners, and some river commissioners were dismissed.

The river basin is a highly integrated and closely interconnected area, in which the upper reaches, middle reaches, lower reaches and different regions influence and constrain each other. This characteristic of the river basin implies that local governments in the region should cooperate with each other to avoid and manage the situation of separate administration. However, due to the strong negative externalities of water pollution in the basin and the strong positive externalities of water environment management in the basin, local governments often choose to "free-ride" on the basis of unequal costs and benefits, resulting in the dilemma of collective action when property rights cannot be clearly defined. To address the problem of inter-administrative cooperation, the comments suggest not only that river chiefs "clarify the responsibilities for managing rivers and lakes in different administrative regions and coordinate the implementation of joint prevention and control measures upstream and downstream, and on the left and right banks," but also that river chiefs supervise the performance of the tasks of the relevant departments and river chiefs at the next level. Under the river chief system, the river chiefs at the higher level can coordinate the horizontal responsibility relationship between the river chiefs at the next level and ensure inter-administrative coordination through stronger evaluation and accountability. Through cooperation among horizontal administrative forces, the river chief system helps local officials take responsibility for water control. Through the administrative integration of higher-level bodies, the special coordination of river basin agencies, and the coordination and linkage between departments, the river basin officer system has formed a "great synergy" water management pattern across levels, regions, and departments.

The River Basin Agency is the water management department established by the Ministry of Water Resources to exercise water management authority on behalf of the Ministry of Water Resources in the river basin under its authority, and to play the role of guidance, coordination, supervision and control in the implementation of the river management system and coordinate the management of the river basin. The "implementation plan" formulated by the Ministry of Water Resources and the Ministry of Environmental Protection clearly requires river basin institutions to actively promote the work of the river chief system, which coordinates the management and protection of rivers and lakes in various places with river basin planning, continuously strengthens the planning constraints of river basin institutions, clarifies river and lake management responsibilities across administrative regions, coordinates upstream and downstream, left bank and right bank, strengthens systematic coordination and control, and implements joint prevention and control. The river basin management agency focuses on coordinating and resolving major issues in the comprehensive implementation of the river chief system in major rivers and lakes, interprovincial rivers and lakes, upstream and downstream, left bank and right bank, and major tributaries, to achieve joint prevention and control and coordinated management, while coordinating issues related to ecological balance in the river basin against the background of the river basin management system. In the process of coordination, the river basin institutions should coordinate the responsibilities of members, assign personnel uniformly, and improve the implementation of the principles of "overall coordination, scientific planning, innovation-oriented and systematic governance".

The Yangtze, Yellow River, Huaihe, Haihe, Pearl River and Songliao Water Management Commissions, Taihu Lake Basin Management Bureau and their subordinate management agencies, as agencies dispatched by the Ministry of Water Resources, guide and promote all works in the basin. This is mainly reflected in the technical support for the implementation of the river head system by guiding the relevant provinces (autonomous regions and municipalities directly under the central government) in the seven major river basins to formulate a coordinated and orderly assessment index system for major rivers and lakes in the provinces; To guide all municipalities to break down the tasks and mandatory indicators of comprehensive river basin planning into regions and river sections; to guide river managers at all levels to implement the requirements of comprehensive river basin planning for rivers and regions at different levels; and to make full use of the leading role of planning in the comprehensive implementation of the river basin system. For example, on December 5, 2020, the Yellow River Bureau participated in the internal review of the key points of the 14th Five-Year Plan for the Protection of the Ecological Water Environment in the Prefectures and Cities in the Yellow River Basin in Henan Province, focusing on the regional strategic positioning and characteristics of the ecological environment, the overall design and planning ideas, etc., technical support for the planning departments of Sanmenxia, Jiyuan, Jiaozuo, Hebi and other prefectures and cities, planning the overall design and ideas, carrying out comprehensive measures and differentiated management, and coordinating and solving trans-regional problems.

Oversight is reflected primarily in the monitoring of the implementation of the various work systems, responsibilities, programs, plans, etc. of the river management system. Supervise and urge all provinces (autonomous regions and municipalities directly under the central government) in the basin of major rivers to solidly promote the tasks of the river basin management system; supervise and urge all municipalities to include the implementation of the plan in the evaluation index system of the river basin management system; supervise the implementation of the overall basin plan, the "three red lines" and the strictest water resource management system in all provinces. The State Council has established a comprehensive river basin coordination committee at provincial and ministerial levels to establish a "1 + 1 + X" coordination and regulation body with "central leadership, local participation and competent river basin agency" for the Yangtze River, Yellow River and other major river and lake basins, whose functions are to implement the policies formulated and decisions made by the Comprehensive River Basin Coordination Committee and to be responsible for guiding, coordinating and supervising matters related to river basin management.

Monitoring is the primary focus of coordinated promotion of river basin management and the river chief system. Specifically, it includes: general monitoring of the current status of water resources, water environment, and ecological security throughout the river basin; monitoring of the carrying capacity of water resources in the provinces within the basin; monitoring of the total amount of pollution across provincial boundaries; monitoring of water quality compliance in water function areas at provincial boundaries; monitoring of indicators in interprovincial water disputes. Taking the Yangtze River Basin as an example, the Yangtze River Basin National Coordination Mechanism coordinates relevant departments of the State Council to improve the monitoring network system and the mechanism for exchanging monitoring information on ecological environment, resources, hydrology, meteorology, navigation and other monitoring systems, natural disasters and other monitoring systems and mechanisms to exchange monitoring information in the Yangtze River Basin based on the established stations and monitoring projects, to obtain relevant information on the ecological environment, natural resources, management and law enforcement in the Yangtze River Basin to ensure that the Yangtze River Basin monitoring system is truly implemented and plays a significant role.

Since environmental protection departments, water management departments, agriculture departments, and fishery departments tend to formulate water-related policies based on their own interests, some policies are in conflict. In addition, the daily work of water control is a separate monitoring and management system, resulting in common problems such as overlapping responsibilities, different authorities and responsibilities, and lack of cooperation between functional departments, which affect the efficiency of water management. In order to improve the performance of interdepartmental cooperative management, the river chief system innovates interdepartmental collaborative linkage without touching the foundation of bureaucratic structure, and fulfills interdepartmental information cooperation and management cooperation.

Cross-sectoral collaboration can be defined as "institutionalized collaborative behavior among multiple actors that transcends organizational boundaries," and this collaboration occurs across different policy areas and administrative regions and is reflected in different levels of decision-making, implementation, and service delivery. Cross-sector collaboration includes both horizontal and vertical collaboration. On the one hand, the river chief system has built a bridge for interaction between departments. On the other hand, the "River Chief System" lays the foundation for integrating the resources of different departments.

After the implementation of the "River Chief System" the river chiefs are empowered to manage and effectively divide the work of all water-related departments in their jurisdiction, avoiding the problem of uncoordinated management and overall coordination of multiple departments. At the same time, the river chiefs have established offices of river chiefs at all levels, which can effectively link horizontal water-related departments such as water conservancy, environmental protection, construction, agriculture, forestry, etc., organizationally ensure the exchange of information and smooth communication between departments, and promote cooperation between departments. This system can integrate the executive power of governments at all levels as much as possible, and through the coordinated distribution of government forces at all levels, it can effectively manage all levels of water environment and effectively reduce the administrative costs and difficulties that may arise from the decentralized administrative structure.

In Qingshan county, Hubei province, a county-level river chief system department linkage work mechanism, and accordingly promulgated the "Qingshan county comprehensive implementation of river chief system work plan", the member units of the county river chief system office and other relevant departments of the county based on the clear division of responsibilities, strengthen communication, exchange and contact cooperation, in the face of interdepartmental important issues, the relevant departments of the county work around the river commissioner system center, according to the division of responsibilities and tasks, strengthen horizontal linkage, close coordination and cooperation, timely exchange of relevant river and lake meteorology, water resources, cross-section water quality, water pollution and other information, and then jointly promote the implementation of the task, and finally establish a standardized, proper, efficient department linkage promotion mechanism. the "River Chiefs" at all levels can not only clarify the responsibilities of different departments and assign water control tasks in a timely manner, but also effectively coordinate and integrate the resources of different departments, implement unified and coordinated management according to the indivisibility of water flow, and improve the efficiency of management.

The "system of river chief system" has established a mechanism for linking upstream and downstream and is able to integrate human and material resources within its jurisdiction to enable coordinated action among the administrative forces of different departments. For example, the Sancha River Basin has adopted a three-tier administrative model with prefecture-level, county-level, and township-level cities, and the staff is composed of leading groups and leading group offices at these three levels. The staff is composed of leading groups and leading group offices at these three levels. A mechanism has been established to liaise between the top and the grassroots, from provincial units to commune heads, and the river basin management situation can be quickly and effectively transmitted between different departments. In the event of an emergency, the river chief can respond quickly and report to the higher-level leaders to guide the relevant departments to solve the river problem in the first place. Another example is that Beijing City has combined the system of "whistling in the streets and communities and reporting to the departments" with the system of river and lake chiefs. As soon as the river chiefs at the street and township level discover problems in rivers and lakes, they send a signal to convene, and all affected departments come on duty. Through this integration of human and material resources, the "system of river chiefs" has evolved into a coordination system that is truly uniformly deployed and jointly implemented.

The public nature of water resources determines that water management is a public matter. Water management is an essential prerequisite for local officials to perform public functions. Against the backdrop of the establishment of the river chief system, local officials have achieved rapid results in water management based on the establishment and operation of a responsibility system composed of political, administrative, legal, and technical responsibilities. Political responsibility is the basic responsibility mechanism of local officials for water management, administrative responsibility is the core responsibility mechanism of local officials for water management, legal responsibility is the basic responsibility mechanism of local officials for water management, and professional responsibility is the basic responsibility mechanism of local officials for water management.

In addition to the bottom-up public supervision, on the one hand, the CPPCC at all levels plays a democratic supervision function in the form of consultation and management of problems that exist in the water management of river chiefs; on the other hand, audit institutions at the county level and above conduct audits of the natural resources of river chiefs at the time when river chiefs leave office, play a supervision and accountability function, and promote the implementation of the people's public will for water management.

The Chinese People's Political Consultative Conference, short for the Chinese People's Political Consultative Conference, is an important institution for bipartisan cooperation and political consultation under the leadership of the Chinese Communist Party. Political consultation, democratic control, and participation in and discussion of state affairs are important functions of the CPPCC. The "Opinions on Strengthening and Improving the Democratic Oversight Work of the CPPCC" issued by the General Office of the CPC Central Committee points out: Democratic supervision of the CPPCC is an important part of China's socialist control system. The National Committee of the Chinese People's Political Consultative Conference and the local committees set up at all levels in various localities are fulfilling their role in participating in state affairs and democratic supervision by convening consultative meetings, making suggestions, supervising members, carrying out research on specific issues, and reflecting social conditions and public opinion on principles and policies as well as on important social realities of concern to the people.

River and lake management is an important matter related to people's production and livelihood, and it is also a concern of the CPPCC. The matters related to people's livelihood should be managed in consultation with the people. The river chief is an institutional innovation that meets the needs of comprehensive management of the water environment in the river basin, and the public has expectations of him. In order to contribute to solving this difficult problem that affects people's lives, the CPPCC has actively taken the initiative at all levels and in all localities, carried out in-depth special investigations and studies on the river basin system, and made suggestions and proposals to the government in the form of proposals, research reports, suggestions and information on social conditions and public opinion. In Fujian, the provincial committee of the Chinese People's Political Consultative Conference focused on the river chief system and held a thematic consultation meeting with the provincial government to address the problems in the operation of the river chief system. In Shaanxi, the Committee on Population, Resources and Environment of the Provincial Committee of the Chinese People's Political Consultative Conference put forward the "Proposal on Problems and Suggestions in the In-depth Implementation of the River Chief System in Our Province" at the second meeting of the 12th Provincial

Committee of the Chinese People's Political Consultative Conference in Shaanxi. In addition, the CPPCC has monitored the implementation of the river basin ladder system in some provinces through on-site inspections. In Jilin, the provincial CPPCC introduces the form of three-level liaison and coordination of the CPPCC to conduct supervision inspections on the implementation of the river head system, and the supervision inspection report prepared according to the actual inspection situation is deliberated and approved by the provincial CPPCC chairman's meeting. In Yunnan, the provincial committee of the Chinese People's Political Consultative Conference led an inspection and supervision team to Qujing to inspect the comprehensive management of water areas on site, listen to relevant information and check relevant accounts; at the same time, the research team of the provincial Ministry of Natural Resources was assigned to supervise the river channel system of the Nu River (Yunnan section) to monitor the discovery of problems and the implementation of corrective measures. In Zhejiang, the provincial CPPCC called on members of CPPCC chairman assemblies of municipalities and counties to serve as river chiefs, and led thousands of CPPCC members to democratically supervise the implementation of the river chief system.

The audit organs established by local people's governments at or above the county level in accordance with national laws shall exercise audit supervision functions and powers on behalf of the State. The 18th Plenary Session of the 18th Central Committee of the Communist Party of China in October 2014 proposed to improve the audit system and ensure the independent exercise of audit supervision rights in accordance with the law. Based on the implementation of pilot projects and the summary of experience, the General Office of the CPC Central Committee and the General Office of the State Council issued the Regulations on the Audit of Natural Resource Assets of Leading Cadres (for Trial Implementation) in December 2017, which clearly regulates the audit of natural resource assets leaving the office of leading cadres and focuses on the quantitative and qualitative changes of natural resource assets and ecological environment in the jurisdiction of leading cadres during their tenure. River chiefs at all levels, as the main managers of inland rivers and lakes in their jurisdictions, are naturally included in the scope of the review of natural resource assets leaving office.

The establishment and implementation of the audit system for outgoing natural resources and the lifetime accountability system for ecological and environmental damage help to avoid blind decisions by leading cadres at all levels without regard to the ecological environment, exclusive concern for economic development without regard to ecological development, and neglect of long-term development by only short-term interests. In the river chief system, local chiefs who serve as river chiefs, whether their term expires or they are promoted, transferred, dismissed, or retired during their term, face independent audits of their natural resources by auditing institutions and lifetime accountability for ecological and environmental damage. These include how well they have protected water resources during their tenure and whether they have adhered to the three red lines of developing water resources and controlling water use, controlling water efficiency and limiting water functional areas, and recording pollution; what progress has been made in riparian management and protection of rivers and lakes; and whether there has been an increase in the area of water and the percentage of natural shorelines to the total area; How well the water pollution prevention and control work is implemented, whether the pollution control on water and shore has been coordinated, and whether the pollution control mechanism and evaluation system are perfect; and the effectiveness of water environment management, ecological restoration of water bodies, and law enforcement and monitoring will be the focus of the audit. Using the river and lake management and protection information system, geographic information system, map GIS and other technologies, and ecological assessment methods introduced by the river administration, the changes of river and lake shorelines (water area and natural shore retention rate), changes in water bodies, water quality, and biology of rivers and lakes, and changes in real-time monitoring data of transboundary sections and key water areas were compared and analyzed and combined with daily inspection logs, problem monitoring and reporting, and complaints. Objective evaluation of the audited river master's performance in managing natural resources and environmental protection during his tenure. The audit results will be an important basis for the evaluation and appointment of the organization's human resources department. Conducting the audit of natural resources leaving office can not only protect the public interests of society, monitor and limit the power, but also help the system of river supervisors to take root and avoid the phenomenon of people leaving the government.

In order to ensure that river managers and competent water authorities fulfill their water control responsibilities, it is necessary to monitor the legality and adequacy of their management activities in accordance with the law. From the point of view of the operation of the power system, supervision is mainly manifested in the supervision of the People's Congress at the same level, the hierarchical supervision of administrative organs, and top-down supervision within the Party. Under the principal-agent relationship, the people delegate public power to the government through the People's Congress, and the relationship between the government and the People's Congress is between the empowerer and the beneficiary. Therefore, the local people's governments at different levels elected by the people's congresses at the same level are the executive organs of the people's congresses at the same level, are responsible to the people's congresses at the same level, and are also subject to the control of the people's congresses at the same level. The local chief of the river, as the representative of public power, should naturally accept the supervision of the person authorized by the People's Congress at the same level.

At present, the Environmental Protection Law of the People's Republic of China, the Law on the Prevention and Control of Water Pollution of the People's Republic of China, and a series of local ordinances and regulations clearly state that local people's governments and river administration chiefs at all levels are responsible for the environmental management of their respective administrative regions. Local Party and government chiefs at all levels who serve as river commissioners are primarily responsible for managing the ecological environment in their jurisdictions. Whether during the People's Congress or during the meeting of the Standing Committee of the People's Congress, local people's congresses at all levels can ask questions to people's governments at the same level about the management of the water environment. Although questioning falls within the scope of consultation, administrative organs shall apply a system of principal responsibility, and principal officers shall be responsible for illegal and neglectful acts at their own level and in their own department in accordance with the principle of consistency of authority and responsibility. In serious cases, they may be removed by the People's Congress at the same level.

In addition to supervising the People's Congress meetings, People's Congress deputies can also urge river officials to carry out work in accordance with relevant requirements through investigation, inspection and supervision, and hold forums to ensure the fulfillment of river and lake management tasks; in some of these areas, individual People's Congress deputies have also been hired as "civilian river officials" to carry out regular supervision of local water management. After Wuhan City in Hubei Province issued Ordinance No. 1 on the "Municipal River and Lake Chiefs System" in 2021, Hu Lishan, director of the Standing Committee of the People's Congress of Wuhan City, led teams to the Zhenghe River Basin twice to check the implementation of the river and lake chiefs system, and held symposiums to listen to the work reports of relevant departments and urge relevant departments to fulfill their duties. In short, with the enhancement of people's awareness of environmental protection and protection of rights, river managers who cause environmental damage due to failure to fulfill their duties will inevitably be examined, questioned and even dealt with by the People's Congress.

The bureaucratic system means not only a solid and orderly system of superiors and subordinates, but there is also a supervisory relationship between superior organs and subordinate organs, that is, hierarchical supervision by administrative organs. In order to improve administrative efficiency, China's Law on Administrative Supervision, Law on Administrative Punishment, Civil Service Law and the "Framework for the Comprehensive Promotion of Law-Based Administration" issued by the State Council have explicitly strengthened the pursuit of responsible administration. In general, administrative bodies, organizations authorized by laws and regulations, organizations entrusted by administrative bodies with the enforcement of laws, and their employees who intentionally or negligently fail to perform prescribed duties or perform them incorrectly and harm the interests of the state and the legal rights and interests of citizens, legal entities and other organizations are investigated for administrative responsibility.

The top-down system of river commissioners is an important step in solving the problem of "water control in Kowloon." It not only affects local interests on the upper and lower reaches of the river basin, on the left and right banks and between the main streams, but also affects the work results of many water-related departments such as environmental protection, land and water management, forestry, agriculture, etc., and also affects and hinders the process of building an ecological civilization. In the process of implementing the river chief system, the central and state ministries and commissions have issued documents repeatedly pointing out the need to strengthen the evaluation and accountability of river chiefs, such as the "Implementation Recommendations for Promoting the River Chief System from "Famous" to "Real"" issued by the Ministry of Water Resources in October 2018, which points out that all localities should strictly implement the assessment of river chiefs at higher levels for river chiefs at lower levels, and use the assessment results as an important reference for the selection and appointment of cadres. It is necessary to establish and improve the accountability mechanism and seriously investigate the responsibility of the relevant river chiefs and relevant river chief departments who do not fulfill their duties, do not act, act slowly and act indiscriminately, so that the outstanding problems of rivers and lakes cannot be solved for a long time.

In implementing the responsibilities of river chiefs, local governments at all levels must formulate annual evaluation methods for the work of the river chief system in accordance with the annual work priorities of the river chief system in their respective regions. The results of the annual evaluation of the river chiefs system at lower levels are used as an important reference for the appointment and dismissal of cadres, the annual comprehensive evaluation, the allocation of project funds, etc. If a river chief at a higher level finds in the daily work that the river chief at a lower level is not performing his duties properly, he can also use methods such as reminders, interviews and criticism reports to hold the river chief at the lower level accountable, and the river chief at the higher level can rely on the office of the river chief system to refer the indications of the problem to the disciplinary review committee and the supervision committee at the same level, and the disciplinary review committee and the supervision committee will handle it according to the rules and discipline.

At the 14th meeting of the Central Leading Group for Comprehensive Deepening of Reform, it was stressed: it is necessary to strengthen the requirements of "equal responsibility of the Party and the government" and "dual responsibility of one body" for environmental protection, and to investigate the responsibility of relevant units and individuals in places where problems arise. The so-called "joint responsibility of the Party and the government" means that the Party Executive Committee and the government at the same level have the same responsibility for environmental protection and supervision. The so-called "one body and two responsibilities" means that Party and government organs and their leaders must not only do a good job in professional work, but also take responsibility for environmental protection management and supervision in the areas for which they are responsible. As part of the comprehensive promotion of the river chief system, a total of 402 provincial-level river and lake chiefs have been appointed nationwide, of which 59 provincial-level Party and government principal officers are river and lake chiefs. The status of principal Party and government officials of each river and lake governor creates the conditions for intra-Party accountability.

Since the 18th National Congress of the Communist Party of China, General Secretary Xi Jinping has repeatedly emphasized, "There must be responsibility for power, responsibility for responsibility, and accountability for failure. In order to unify and strengthen the Party's accountability, the Party Central Committee has formulated the Regulations on the Accountability of the Communist Party of China, which clearly states that "in promoting "the construction of ecological civilization, major deviations and mistakes occur, causing serious harm to the Party's cause and the interests of the people" Party organizations and leading Party cadres will be held accountable, and this accountability is a lifelong accountability, and "for the serious nature of the dereliction of duty and the grave consequences, whether the person responsible is transferred, promoted or retired, he or she must be held seriously accountable." Therefore, under the system of the Chinese Communist Party leading the government, any Party or government official who acts as a river commissioner will be comprehensively evaluated and held accountable by the higher Party organization. From top to bottom, if a water accident occurs or water control objectives are not met in a particular location, the local chief executive as the first in charge of the government system and the Party committee chairman as the first in charge of the Party committee system will be held administratively, politically and disciplinarily accountable.

After the Central Eighth Ecological Environmental Protection Supervision Group reported in April 2021 that the city of Baoshan in Yunnan Province had failed to control pollution and the eastern river of the 'Mother River' had become a polluted river", the Yunnan Provincial Discipline Inspection Commission and the Yunnan Provincial Construction Commission launched an investigation on the instructions of the Secretary of the Provincial Party Committee and the Governor, and based on the results of the investigation, in accordance with the regulations on accountability of the Communist Party of China, the Regulations of the Communist Party of China on Disciplinary Measures of the Law of the People's Republic of China on State Sanctions Against State Officials and the Measures for Prosecuting the Responsibility of Leading Party and Government Cadres for Ecological and Environmental Damage (for trial implementation), and other relevant Party disciplinary laws and regulations, notified and held accountable four responsible units and seven provincial-level cadres, 12 department-level cadres and two section-level cadres admonished and held accountable, criticized and enlightened, registered in government affairs, registered in important government affairs, warned internally within the Party and seriously warned. In Fuxi Municipality, Huizhou County, Huangshan City, the combination of the river chief system and rural party building has produced a new mechanism "party building + river chief system". Fuxi Township requires each Party member to ensure the management of a river section, strengthen the pioneering and exemplary role of communist Party members, and set an example of Party members' responsibility on the river. Party branches in all villages actively carried out Party activities on the "river chief system" to spread knowledge of river and lake protection among villagers along the route. The aim is to ensure that no flotsam on the river surface, no obstacles in the river and no exposed garbage remain on the river bank, and to achieve the ecological goal of "smooth river, clean water, green bank and beautiful landscape".

2 Forms of Responsibility

Political responsibility is the fundamental responsibility of modern democratic politics, which arises from responding to the demands of the public will (Han & Li, 2021). In the face of increasingly complex and deteriorating ecological conditions of water bodies, ecological water management has become a demand of the public will of the people in the new era. Local officials at all levels, as those responsible for implementing the people's public will, guide water conservation and implement political responsibility guided by the promotion of public interests. Political accountability has also become the basic accountability mechanism for local officials in water management.

In all government systems, there are two main or basic functions of government, namely the expression of the national will and the executive function of the national will, that is, political and administrative. The will of the state is the will and aspiration of the state for the ruling class in a country to raise the political, economic and other demands of their class to the will and aspiration of the state through legislation in order to safeguard the interests of their own class. Politics is also the concentration and expression of the public will of the broad masses of the people.

In order to realize the will that the state wants to express and form it into a code of conduct, this function must be placed under political control. Therefore, governments at all levels must assume appropriate political responsibility in the exercise of their administrative functions. Political accountability is the fundamental task of modern democratic politics. In modern countries, depending on factors such as population size, population quality, and territorial extent, no form of political power organization can guarantee the direct exercise of state power by all the people of a country. The representative system has become an important means of exercising political responsibility. Under the representative system, the public delegates its power to the government, and the government exercises public power on behalf of the public, and the public obeys the public power with the expectation that the public power can fulfill its interests or demands This means that the government must respond to and fulfill the demands of the public will of the public in order to fulfill its political responsibility.

Since ancient times, China has faced the basic national conditions of extremely uneven spatial and temporal distribution of water resources and frequent floods and droughts, forcing the government to undertake public flood control tasks. In ancient China, large-scale centralized water control by the central government was not the norm, while local government on one side of the government was the norm for regional water control. Especially in the middle and lower reaches of the Yellow River, the middle and lower reaches of the Yangtze River, and areas with frequent flooding in the southwestern region, the implementation of the public function of water regulation became the main task of the local chief official to "serve as an official and benefit a party." In Chinese history, there were also many cases where local governors successfully controlled water, such as Li Bing, the commander of Shu County in the Qin State who built Dujiangyan, the Su Dongpo of Hangzhou in the Song Dynasty, Yu Qian, the governor of Henan and Shanxi Provinces in the Ming Dynasty, and the Yellow River in Jiangsu in the Qing Dynasty. These local governors use limited social resources to maximize the transformation of floods into water conservation, promoting local social stability and economic development.

With the transformation of traditional countries into modern countries, the public management functions of the government have been continuously strengthened and refined in content, but the public functions of water control have never been forgotten. After the establishment of the People's Republic of China, under the cry of "We must repair the Huai River," a boom in the construction of water conservancy facilities was launched throughout the country. During collectivization, more than 80,000 large and medium-sized reservoirs were built throughout the country, and the irrigated area accounted for more than 50 percent of the total agricultural land, a 3.5-fold increase over the irrigated area before 1949. Although there was blind construction and even environmental destruction in the construction of reservoirs during the collectivization period, these reservoirs effectively ensured water supply for agricultural production and greatly improved agricultural productivity.

After reform and opening up, economic construction has become the focus of the Chinese government's work. However, due to a lack of environmental awareness, the rapid development of the economy in various places has also led to problems such as water pollution, damage to the water environment and degradation of water ecology, which seriously threatens water security. In many places, the old problems of water ecology have not yet been solved, and new problems have gradually emerged, which are even extremely harmful. Therefore, General Secretary Xi Jinping stressed that "water security is an important matter related to the country's long-term peace and stability," and that "we must protect the ecological environment like our eyes and treat the ecological environment like life." Natural water resources that belong to the public are naturally competitive and non-exclusive. Even in the face of intensified ecological water management, the nature of public affairs itself still leads to the tragedy of the commons, the prisoner's dilemma, the "collective action problem," and even "a group of helpless individuals caught in the cruel process of destroying their own resources." Although the "self-organised governance model" proposed by Ostrom effectively circumvents the dilemma of the "national coercive plan" and the "private property rights plan" for water management, it is difficult to be effective in the face of large-scale problems with natural water resources. Therefore, in the face of increasingly complex and deteriorating ecological conditions in the water sector, the government is urgently required to respond to public opinion and perform the public function of water control.

In China, the People's Congress system is the basic political system of our country. The people's congresses at all levels, democratically elected by the people, are the organs of state power in our country. The people's governments at all levels are elected by and accountable to the people's congresses at the same level; the heads of local people's governments at all levels report on their work to the people's congresses at the same level and are under their supervision. The "Opinions on the Comprehensive Implementation of the System of River Chiefs" issued jointly by the General Office of the CPC Central Committee and the General Office of the State Council points out that each province (autonomous region or municipality directly under the central government) should appoint a general river chief who should be the chief comrade in charge of the Party Committee or the government; In each province (autonomous region or municipality directly under the central government), river chiefs shall be established on the main rivers and lakes in the administrative regions, with responsible comrades at the provincial level acting as responsible comrades; in cities, counties and municipalities where rivers and lakes are located, river chiefs shall be established at different levels and in different sections, with responsible comrades at the same level acting as river chiefs. Therefore, local chiefs at all levels as river chiefs are representatives of public power responsible for responding to and implementing the public will of the people, and of course the main body of political responsibility.

Public opinion is different from public opinion, "public will focuses only on public interest, while public opinion focuses on private interest" In view of the demand of people's public will for ecological water management in the new era, river basin supervisors at all levels take the maintenance and expansion or reduction of public interest as the starting point and basis for the implementation of water management. In the process of introducing the river basin supervisor system, the main local party and government cadres at all levels have significantly raised their awareness of environmental protection, and while striving to develop the economy, they have consciously and comprehensively considered the pollution problem and no longer regard economic growth as the only development goal as in the past. When it comes to attracting investment, we can boldly say no to production projects with high energy consumption, high pollution and low added value. As one district leader said, "if a polluting project is introduced, the environment will definitely be polluted, the lives of ordinary people will be affected, and there must be opinions about the government. At the end of the day, I, the head of the river, must regulate it".

The administrative responsibility that results from the implementation of the "public will" of the people is the central mechanism of responsibility of local officials for water management. In the process of comprehensively introducing the river basin manager system, China has established a systematic and perfect organization system for river basin managers, in which the administrative responsibilities of river basin managers are clearly defined at all levels and in all sections, and the river basin manager leads the implementation of the public responsibilities of water management to promote the implementation of the people's "public will". The administration is the implementation of the will of the state. For a government that has both political and administrative functions, the expression and implementation of the will of the state are essentially "two sides of the same coin." This is because "the conduct of the state as a political entity consists both in the activities necessary for the expression of its will and in the activities necessary for the execution of its will." Local party and government leaders at all levels, while responding to the "public will" of the people, must lead and represent the local government to implement the "public will" of the people and carry out the administrative tasks. It can be said that the process of implementing the "public will" by local officials is that they are responsible for and perform administrative duties to their entitled subjects, the public.

"To administer the country responsibly and conscientiously" is the basic requirement for every local official. In the implementation of the "public will"," local officials are guided by factors such as political performance and local interests, resulting in "localism" such as "local protectionism" and "localism" Before the introduction of the river basin ladder system, under the influence of "localism", there was a contradiction between the integrity of the ecological environment of the basin and the fragmentation of administrative areas in the field of water control, and the integrity of the ecological environment of the basin required the management of the water environment of the basin to follow the system concept and realize supra-regionally coordinated water control. The fragmentation of administrative areas has led to party and government leaders at all levels and in all localities giving priority to local interests in water management, which makes it difficult to achieve cross-regional cooperation. At the same administrative level, "departmentalism" has led to "fragmentation" of water pollution control among different government departments, although the departments of development and reform, environmental protection, water conservancy, agriculture, forestry and other departments have done their own work, but because of the lack of holistic administrative thinking and horizontal cooperation, the results of segregated water control and the problem is still in the water while the root is still on the shore. In the new era, in the face of public opinion on ecological water management, China has adopted a river head system, in which local heads act as river heads to carry out the public tasks of water management, and administrative responsibility has become the central responsibility mechanism of local heads for water management.

The responsibility of the administration depends on the existence and functioning of the system of administrative organization of the government. As part of the comprehensive implementation of the river chiefs system, all provinces, autonomous regions and municipalities directly under the central government have a four-tier river chiefs organization system at the provincial, municipal, Counties and cities within the province in accordance with the requirements of the opinions, and some provinces (autonomous regions and municipalities directly under the central government) have also extended the river chief organization system to villages (communes) to establish a five-tier river chief organization system. Each local government has appointed a general river chief at the provincial, city, county and township levels, with the party and government chiefs at the same level and having primary responsibility for water control in their jurisdictions; Second, according to the vertical administrative level, a general river officer, a deputy river officer and a river officer at different levels shall be established, staffed by the Party's and the government's principal officers and the principal officers at the same level, and shall have the specific responsibility for water control in the province according to the administrative level; Third, river chiefs shall be established at the provincial, municipal and county levels for the major rivers in their jurisdiction as the river basin unit, and the chief officers of the Party Committee, the government, the People's Congress and the CPPCC and the chief officers of the member units of the river chief system at the same level shall take responsibility for coordinating the management of the river basin.

The establishment and improvement of the river chiefs' organizational system provides the basis for clarifying the administrative responsibilities of river chiefs at all levels and sections. Through the holistic and systematic design of the top level, the river basin supervisors' system has created a top-down, layered and interlocking system of administrative responsibility ""from the whole to the individual". In addition, some local governments have, in practice, detailed the administrative responsibilities of river superintendents by subdividing jurisdictional areas and creating lists of responsibilities to ensure that responsibilities are assigned to individuals. This "vertical to the end, horizontal to the edge" system of administrative responsibility provides a basis for river managers to carry out their responsibilities. For example, many rivers and lakes in the province span various counties, cities and towns, creating a dilemma between upstream and downstream, and between the left and right banks. The provincial-level river commissioners, city-level river commissioners and county-level river commissioners have overcome the "limited" water control space of the main party and city, county and municipal government, and put the water areas at the municipal, county and city levels into a larger space for management, overcoming the "limited rationality" of water control by a single local unit.

Local party and government leaders are not only the representatives of the local public will, but also the primary agents of local public power and the individuals best able to mobilize public resources. Under the bureaucratic system, local officials use the principle and inertia of the bureaucratic system that "the top" is responsible" to dissect and reduce the task of water control layer by layer, creating a "pressure" effect" and improving the awareness of subordinate administrative leaders. Local party and government leaders use political mobilization techniques such as organizational mobilization and emotional mobilization to arouse the enthusiasm of lower-level party and government leaders for water management; at the same time, river directors at all levels use the "river director's office" platform to break down the communication barriers between water-related departments and strengthen the joint force of systematically controlling various departments, divisions and links. Overall, a working pattern should be formed in which "each level is grasped and the implementation at each level" is grasped".

Under the constraint of administrative responsibility, local party and government leaders acting as river managers will not only lack a sense of responsibility because they are far away from the rivers and lakes, but will actively exercise their own initiative, integrate various administrative resources, creatively solve river ecology problems, and respond to the people's "public opinion" in a timely manner. In particular, as the main leader of the local party and government, under the guidance of the systematic water management thinking, which regards the landscape and forest-field-lake as a community of life, he coordinated water management with mountain control, forest management and field management, jumped out of the trap of "point-point water control" and "nine-dragon water control", and introduced "total water control" and "centralized water control", which not only improved the utilization rate of water resources, but also solved the problem of cooperation between upstream and downstream and left bank and right bank.

As an integral part of building an ecological civilization, the management and protection of rivers and lakes is naturally an important part of national governance. In order to govern the country according to law, water must be managed according to law. In establishing the system of river chiefs, China defines the responsibility of local chiefs who act as river chiefs for water management at the three levels of national laws, Party discipline and Party rules, and local regulations, clarifies the basic responsibility of local chiefs for water management, and ensures long-term "river chief governance" through water management according to law. Law is the most basic and fundamental embodiment of the "public will" of the public. As the representative and executor of the "public will" of the people, the government must abide by the relevant laws and bear the appropriate legal responsibility in implementing the "public will." Plenary Session of the 18th. The decision of the Central Committee of the Communist Party of China on some important issues of the comprehensive promotion of the rule of law adopted at the 18th Plenary Session of the 18th Central Committee of the Communist Party of China pointed out that in order to comprehensively promote the rule of law, it is necessary to adhere to the joint promotion of the government of the country according to law, it is necessary to adhere to the joint promotion of the government of the country according to law, government by law and administration by law, to adhere to the integrated construction of a country by law, government by law and society by law, to achieve scientific legislation, strict law enforcement, fair justice and law-abidingness of the whole people, and to promote the modernization of the national government system and government capability. As an important part of building an ecological civilization, water management is naturally an important part of national governance. When they are in charge of water control, local chiefs who are river chiefs must abide by the relevant provisions of water control laws and regulations and bear the legal responsibility for water control.

Unlike political responsibility and administrative responsibility, the content, scope and time limit of legal responsibility are clearly defined, and the implementation of legal responsibility is supported by the state's force apparatus, which is obligatory and irresistible by the state. For the responsible entity, legal responsibility is an insurmountable limit that must be fulfilled. At the beginning of its introduction, the river basin supervisor system was challenged by the "strong color of human rule", and some scholars believe that this way of managing rivers and lakes, which emphasizes human rule over legal rule, is not a long-term environmental management model. With the nationwide implementation of the river chiefs' system, the basic responsibility of river chiefs for water management has been established in laws and regulations at all levels, creating a long-term and stable constraint for river chiefs at all levels and avoiding the short-sightedness, arbitrariness and contingency of river chiefs in water management. At the same time, river chiefs are sanctioned by laws and regulations when they fail to fulfill their legal duties, which are a violation of the laws and regulations, and this negative incentive pushes river chiefs at all levels to manage water in accordance with the law.

The river chief system is not only a policy but also a system of comprehensive water management and a general term for legal norms concerning the management of rivers and lakes. In the course of gradually promoting the river chief system, China has continuously improved the laws and regulations implementing the river basin management system at the national and local levels, defined the responsibility of river basin managers for water management at the legal level, and urged river chiefs to manage water in accordance with the law by assuming legal responsibility. At the national level, Article 6 of the Environmental Protection Law of the People's Republic of China, which came into effect on January 1, 2015, stipulates that the responsibility for the environmental quality of administrative areas at all levels lies with the respective people's governments. As legal representatives of the people's governments, local administrators at all levels are naturally responsible for the environmental management of the relevant administrative regions. If the local people's governments unilaterally attach importance to economic development and neglect environmental protection, they are violating the law. In order to further clarify and strengthen the responsibility of Party and government leaders of local governments at all levels for water management, the Law of the People's Republic of China on the Prevention and Control of Water Pollution, amended in 2017, incorporated the "system of river chief system" into the legal framework, clearly defining the organizational structure at the provincial, municipal, county and city levels, as well as the six main areas and specific work contents for which river managers are responsible. Local party and government leaders at all levels who serve as river officers have legal responsibility for protecting the water environment in their administrative areas.

Second, within the framework of Party discipline and Party regulations, the Organization Department of the Central Committee and the Ministry of Supervision have issued and implemented the "Measures for Investigating the Responsibility of Leading Party and Government Cadres for Ecological Environmental Damage (Process)" in accordance with relevant internal Party regulations and national laws and regulations, to strengthen the responsibility and competence of leading Party and government cadres for protecting the ecological environment and resources, it clearly proposes to implement the "joint responsibility of the Party and the government" in relation to the ecological environment and establish a system of lifetime accountability for leading cadres who cause serious damage to the ecological environment. All Party and government leaders who cause serious damage to resources or the ecological environment regardless of the laws of scientific development will be held strictly accountable, regardless of whether they are transferred, promoted or even retired. In 2017, the General Office of the CPC Central Committee and the General Office of the State Council issued the Regulations for the Natural Resource Audit of Leading Cadres Leaving Post (for Trial Implementation), which stipulates that in addition to auditing the work related to building ecological civilization and protecting natural resources of leading cadres, "compliance with laws and regulations on natural resource management and ecological environmental protection" must also be audited.

Finally, at the level of local regulations, local governments at the provincial level enact specific laws based on the actual situation of local implementation of the river chief system. For example, Zhejiang, Hainan and Jiangxi provinces have successively formulated and promulgated the "Regulations on the River Chief System of Zhejiang Province", the "Regulations on the River Chief System and the Lake Chief System in

Hainan Province" and the "Regulations of Jiangxi Province on the Implementation of the River Chief System and the Lake Chief System" to ensure that the river chief system works in a legal way. In terms of specific content, for example, the Regulations on the River Chief System of Zhejiang Province clearly stipulates the responsibilities of river head system working organs at or above the county level and the penalties for violations, the specific working responsibilities of river chiefs at all levels (provinces, cities, counties, townships and villages) and the penalties for violations, and the methods, channels and effectiveness of public participation in the river chief system. It not only sets the "baseline" for river chiefs at all levels to carry out their duties, but also shows the direction.

The fundamental characteristic of law is stability. Establishing river chiefs' responsibility for water management in the form of laws and regulations not only provides legal guidance to local chiefs acting as river chiefs on how to implement water management and protection in practice, but also provides a legal basis for investigations against river chiefs for breaches of duty. First, improving laws at the national level provides a legal basis for local officials to introduce and implement the river chief system in their administrative regions. Second, since the local chief official is not only the head of the local administration but also a member of the Communist Party of China, internal Party regulations provide for the accountability of leading cadres for ecological and environmental damage and the auditing of natural resource assets after leaving office, and together with national laws form a "double insurance" that encourages local officials to fulfill their responsibilities in managing and protecting water resources in accordance with laws and regulations. Finally, the provincial-level administrative regulations formulated by each province for the river commissioner system clearly define the responsibilities of river commissioners at all levels in the administrative region, and provide action instructions to river commissioners at all levels for the ecological management of rivers and lakes. It can be said that establishing the legal responsibilities of river chiefs in the management and protection of rivers and lakes in the form of laws and regulations, and promoting river chiefs in the management of water under the law, is a long-term mechanism to ensure the realization of "river chief governance".

The complexity of managing and protecting the water environment means that water management must be based on professional knowledge, which means that river chiefs must take professional responsibility for water management. In practice, the local chiefs who act as river chiefs involve professional forces inside and outside the system to ensure professional water management, so that water management is effective. In the traditional period, the three main tasks of water control in China were flood control, water transportation and irrigation. With the rapid development of the modern economy, people's use of water resources has become more diverse, and the accompanying water pollution, ecological degradation of water bodies and shoreline degradation have become new problems facing water control. In this context, water control concerns not only upstream and downstream, left bank and right bank, but also the bottom, water body and water surface. It is a long-term and systematic project that requires comprehensive management to simultaneously solve the problems of water safety, water resources, water environment, water ecology, water landscape, water culture and other aspects.

The "Opinions on the Comprehensive Implementation of the River Chief System" issued by the General Office of the CPC Central Committee and the General Office of the State Council pointed out: In implementing the river chief system, all municipalities should adopt a problem-oriented approach, adapt measures to local conditions, be guided by the actual conditions of various rivers and lakes in different regions, make overall plans for the upper reaches and lower reaches and for the left and right banks, implement a policy for one river and a policy for each lake, and properly solve the conspicuous problems in the management and protection of rivers and lakes. China has a vast territory, developed water systems, and the conditions of different rivers (lakes) are very different. "One river, one policy" and "one river and one gear" are the premise and foundation of river and lake management. From the perspective of the six main tasks of river chiefs at all levels to carry out water management, whether it is the protection and treatment of water resources and water environment, the prevention and restoration of water pollution and water ecology, or the proposal of feasible plans such as the management and protection of the banks of rivers and lakes and the supervision of law enforcement, professionals must make scientific decisions based on local conditions. Scientific water management strategies must be analyzed, studied, and planned based on a comprehensive understanding of the history and current situation of water quality, quantity, hydrology, energy, and environment in the waters under their jurisdiction.

The complexity of water management makes it necessary for river managers at all levels to rely on professional forces that provide professional technologies and expertise to carry out professional water management and fulfill their responsibilities, whether from the perspective of the general management of today's water environment or from the specific management of each individual river (lake). Professional responsibility has become the fundamental task of river supervisors to manage water effectively. Although local managers are not necessarily experts in water management, they are often the "first person" in their management area, mobilizing various public resources. In water management practice, local leaders take the lead role in overall planning and coordination, engage professional forces within and outside the administrative system to carry out professional water management, and perform professional water management tasks.

On the one hand, local officials coordinate professional forces within the government and strengthen the design of water conservation at the highest level. The complexity of water control means that in the implementation of water control, the relevant departments of the government dealing with water must coordinate water control. Through the river chiefs' working conference, joint meeting of river chiefs, river chiefs' information exchange platform, etc., the local chief official who acts as the river chief involves professional officials from relevant departments such as development and reform, industry and information, finance, planning, transportation, environmental protection, land, municipal government, agriculture, water conservancy, housing and construction, etc. to carry out high-level planning to achieve the overall goal of river and lake management and protection, and make full use of the professional advantages of each department under the premise of holistic management. Promote the transformation of water control methods from "spot" to "integrated" and from "decentralized" to "centralized", gradually realize the specialization, standardization and upgrading of river and lake management and protection, and establish a professional foundation for the effectiveness of the "river chief system".

On the other hand, local officials recruit professionals from outside the administrative system to participate in water control and provide professional support for water control. To compensate for the lack of professionals and expertise within the system, local governments at all levels actively cooperate with universities, scientific research institutes and other units, and bring relevant water professionals into the preparation of local water management plans to provide expertise for local water management. At the same time, in addition to government river chiefs, various localities have established various types of civilian river chiefs, such as "river chiefs", "cycling river chiefs", "postman river chiefs", "river guards of rivers and lakes" and "yang river chiefs", actively attracting civilian river chiefs to participate in water control, to take full advantage of the professional research and "local expertise" of private river chiefs in the practice of water control, so that private river wardens can become effective assistants to official river chiefs and promote the success of water control.

3 Voluntary Being Accountable

Representativeness and accountability are core values of democracy, and public accountability is its due. However, any bureaucratic administration tends to exclude the public and use its advantages to conceal as much information as possible to avoid criticism of its actions and public officials, particularly those who have successfully survived politically, are better at evading responsibility and providing explanations for unpopular actions or decisions in order to avoid being blamed, they even used their professional knowledge to systematically mislead legislators and come to the opposite conclusion. In the political practice of various countries, not only have concrete technologies for avoiding responsibility, such as limiting the agenda, re-setting the agenda, shifting blame among each other, and inertia of behavior, come into being, and has formed the organization strategy, the appearance strategy and the policy strategy and so on avoid responsibility strategy and also has the dynamic choice avoid responsibility model. Chinese government officials, like those in the west, have strong incentives to avoid blame and punishment. Among them, blocking information channels, evading responsibility (looking for scapegoats), mobilizing social networks are its common means. With the promotion of local government reform and innovation, the phenomenon of shirking responsibility has not only appeared from lower officials to higher authorities, but also resulted in inter-governmental cooperation. However, in the process of the implementation of the river long system, the local government (chief officer) not only did not evade responsibility but invited social supervision on its own initiative. Wujiang district of Suzhou City, on the basis of the public notice board, has made the electronic ID card for the river and lake with the help of "Internet +", which is convenient for people to supervise the progress of water control Congjiang County in Guizhou province independently developed the new media platform of "River Patrol Treasure", to facilitate the local people to monitor the performance of the river; Tianjin, Yunnan, Dali Bai Autonomous Prefecture, Shanghai Minxing District actively recruited social supervisors of the river system and even set up social supervision groups to voluntarily accept social supervision and accountability. This kind of self-imposed accountability—beyond what the law requires—is similarly counterintuitive in other countries, and Western scholars have Exploratory research that there are two logical reasons for this.

First, the appropriate logic of moral promotion. Accountability is not just a mechanism, it is a virtue under the logic of propriety that governmental organizations and magistrates "follow the rules which are considered by society to be normal, true, correct or good, without regard for consequences or intended utility" take the initiative to accept accountability, or even ask for social accountability. Mudgee and Morten Olsen found that the expectation of legitimacy through reciprocal, mutually controlled delegation of authority was not sufficient to address the difficulty of spying between client and agent. Trust, conditional on reciprocity, is only an extension of politics. In fact, trust associated with the empowerment of political institutions, like the division of labor itself, is not a clear contract but a rule of proper conduct. The logic behind such appropriate behaviour is essentially a moral constraint on behaviour imposed by values. In Nozick's view, moral initiative is mainly reflected in two aspects: moral thrust and moral pull. The so-called moral thrust, refers to the values as a personal act of moral constraints, to form a "good person" or a moral person's impulse to achieve a person's value. The so-called moral pulling power refers to the values mold individual to other people's moral behavior and respects other people value. Voluntary ethical acts of accountability occur only when the moral force is greater than or equal to the moral force. Dubnik sees morality as a system, and studies it from four dimensions: legal, organizational, professional, and political, the legal system under the accountability system, the responsiveness of government organizations, the sense of responsibility of civil servants, and the political responsiveness are merely manifestations of moral pull, the internalization of legal consciousness, the initiative response of government organizations, the loyalty of civil servants and the initiative response of politics constitute the moral impetus (Dubnik, 2005). Voluntary accountability resulting from the combination of moral thrust and moral pull is the result of the "logic of appropriateness".

The second is the consequence logic under the threat of legislation. In addition to the logic of appropriateness, Koop argues that public organizations choose voluntary accountability in response to legislative threats and to measure differences in the degree of voluntariness of accountability in public organizations and used voluntary accountability data from 103 independent institutions in the Netherlands, using a linear regression least square model and an ordered logit model, the results show that: (1) the more public organizations deal with political issues, the more they tend to commit to voluntary accountability, in order to pre-empt more laws and regulations;

(2) public organizations that provide public services are more responsive to public feedback and more willing to be held accountable than public organizations that are responsible for oversight and management, this is partly to bridge the gap between providers and consumers of public services, and partly because of pressure from politicians for a timely response to public accountability; (3) given the legitimacy, efficiency and effectiveness requirements for the use of public funds, public organizations that rely on government funding are necessarily more pro-active in their accountability than public organizations that rely on private sector funding; Politicians, who are responsible for the use of public money, are also more concerned about the financial management and accountability of government-funded organisations; (4) although public organizations established under private law have public powers similar to those of public organizations established under public law when they are authorized by the government, the public does not recognize them as governmental organizations. At the same time, politicians argue that public organizations established under public law are more liable than private law organizations; and (5) the longer a public organization is in existence, the less necessary and motivated it is to provide voluntary accountability; for young public organizations, voluntary accountability is even more necessary (Koop, 2014).

However, in the view of the principal and agent theory, the accountability system established to avoid the risk of the agent unfaithfully pursuing the principal's interests, regardless of the accountability method (compulsory or voluntary) used, will have a negative effect on the agent. The empirical research on voluntary accountability is insufficient for the real motivation of agents under the negative influence of accountability. In modern democratic society, countries and governments around the world have formed the consensus of "being responsible for the people" and made a commitment to it. To fulfil this commitment, governments have adopted different democratic systems, forms and instruments. The willingness of governments that pursue good governance to provide information and explanations of their actions to the population without being held accountable and subject to sanctions for such actions is an inherent requirement of the development of People's Democratic systems.

Whether it is the "rational virtue" cultivated by long-term school education, or the "ethical virtue" formed by long-term edification of social customs and habits, government officials, including Communist Party members, agree on the basic requirements of the Constitution of the Communist Party of China and the Civil service of the People's Republic of China for party members and civil servants to serve the people heart and soul. Especially as a public servant, it is not only an impulse to realize moral purpose under the rational guidance, but also a rational impulse to respond to ethical requirements, to display personal value and to gain the approval of others. For public officials who act in accordance with the law, the professional code and the requirements of their functions and powers, it is necessary for them to shoulder their responsibilities and actively respond to public queries and accountability in order to fulfill their basic duties. Because political responsibility, obligation, and loyalty are more specific than any commitment, and may not even be voluntary.

With the continuous promotion of the river chief system, many places have creatively implemented the two kinds of river chief system. While the governors, mayors, county heads and township (village) heads are the river heads, the main leaders of the party organizations at the same level are the river heads, thus realizing the joint efforts of the party and political water. In the view of leadership theorists, leaders should not only realize social approval of themselves, but also pursue social approval of their teams. This dual pursuit will certainly prompt leaders to take an active interest in the needs of the people and respond in a timely manner. This shows that both government officials, as well as the principal officials of the government or the party committees, based on their identity and role, and under moral constraints and incentives, are able to generate a voluntary incentive to be held accountable.

The Opinion of River Chief System were issued on 28 November 2016 and were fully completed by the end of June 2018 in just 19 months. River long system can be built in such a short period of time and achieved results,and cannot be separated from the CPC Central Committee and the State Council's high-level promotion. That is to say, under the promotion of the political potential energy of the highest rank, and under the stimulation of multiple incentive mechanisms, the local party committees and governments at all levels not only make clear the responsibility of the main body, but also stimulate the main body consciousness. Under the pressure system and the promotion tournament system, no local party or government leader is willing to put his job in a passive position. Especially under the conditions of moral impetus and promotion pull, local party and government leaders are more willing to cooperate with the party and the country to implement the strategy of ecological civilization construction. For example, out of 402 provincial governors of rivers and lakes in the country's 31 provinces (including the Xinjiang Construction Corps), 59 provincial leaders of the party and government are the general governors of rivers and lakes. In addition to the four provinces (autonomous region and municipality directly under the central government) of Shanghai, Tianjin, Yunnan and Inner Mongolia, the party secretary of the province (region) or the head of the province (City) shall be the chief of the river, the other provinces are both provincial party secretaries and governors of the Shuanghe River. In order to better promote and implement the river length system, the People's Congress and Jiangxi province have taken the lead in issuing the regulations on the river length system for lakes, Jilin, Fujian, Liaoning and other provinces (municipalities directly under the central government) have also successively introduced special legislation on the system of river chiefs, which not only greatly ensures the performance of the duties of river chiefs at all levels in their respective administrative regions, and it largely evades the possibility of passive accountability.

In order to promote the work of the river long system, local party committees and governments have actively set up the "river chief system leading group" in response to the power and influence, and have senior cadres as office directors. Such as Guangdong, Sichuan, Jiangxi, Liaoning, Hunan and other provincial office director, there are deputy provincial leading cadres as. In Jiangxi province, the director of the provincial governor's office is the vice-governor of the provincial government, and the deputy director-general of the provincial government is the deputy secretary-general and the director of the Water Resources Department, the leaders of the Department of Agriculture and Industry, the Department of Environmental Protection, the Department of Housing and construction, the Department of Agriculture, the Department of Forestry, the Department of Transportation, the Department of Industry and information, the Department of land and resources, the Department of Public Security and other departments of the provincial Party Committee shall serve as deputy directors. In terms of the composition of the director-general's office in Jiangxi province, the director of the deputy provincial office is more "powerful" than the director of the office at the department level (the director of the director-general's office in some provinces is the director of the Water Resources Department), it not only helps to solve the problem of pollution in water and roots on shore, but also helps to realize the systematic governance of "the life community of mountains, rivers, forests, fields and lakes. By emphasizing and promoting the responsibility of working groups and team members, for their own in the face of accountability to create more room for maneuver.

Reference

Dubnik, M. (2005). Accountability and the promise of performance: In search of the mechanisms. *Public Performance & Management Review, 28*(3), 376–417. https://doi.org/10.1080/15309576.2005.11051839

Han, Z., & Li, C. (2021). How responsibility is constructed: River chief system and its implementation in S city. *Public Management Research (in Chinese), 44*(1), 137–145.

Koop, C. (2014). Theorizing and explaining voluntary accountability. *92*(3), 565–581. https://doi.org/https://doi.org/10.1111/padm.12058

Chapter 6
Conclusion: The Future of River Chief System

1 Responsibility System and Water Management

Against the background of the establishment of the River Chief System, local officials have achieved rapid results in water management based on the establishment of a responsibility system composed of political, administrative, legal and professional responsibilities. Political responsibility is the basic responsibility mechanism of local officials for water management, administrative responsibility is the core responsibility mechanism of local officials for water management, legal responsibility is the basic responsibility mechanism of local officials for water management, and professional responsibility is the basic responsibility mechanism of local officials for water management.

The first is to take political responsibility for ecological water management. The political responsibility of local officials is based on responding to the "public opinion" of the people. In the face of increasingly complex and deteriorating ecological water conditions, ecological water management has become a requirement of the people's "public will" in the new era, which makes it urgent for local government officials to seriously implement the political responsibility for ecological water management in the process of water management, guided by the promotion of public interests, which has also become the basic responsibility mechanism of local government officials for water management.

The second is the administrative responsibility to implement the public will for water management. The implementation of the people's demand for "public will" has led local government leaders to take responsibility for water management as river chiefs. In the face of public opinion on ecological water management, China has established a system of river chiefs, improved the organizational structure and responsibility system of river chiefs for water management, and encouraged river chiefs to actively lead the implementation of public tasks of water management, which has also become the core responsibility mechanism of local officials for water management.

Y. Hao and T. Wan, *The River Chief System and An Ecological Initiative for Public Participation in China*, https://doi.org/10.1007/978-981-99-4921-2_6

Third, the legal responsibility for water management is carried out in accordance with the law. The profound development of the rule of law has set out the requirements for water management in accordance with the law and provided a legal basis for local officials to carry out their water management responsibilities in accordance with the law. In the process of introducing the River Chief System, China defines the responsibilities of local chiefs who act as river chiefs for water management at the three levels of national laws, Party discipline and Party rules, and local regulations, clarifies the basic responsibilities of local chiefs for water management, and ensures long-term "river chief governance" through water management in accordance with the law.

The fourth point is to strengthen the technical responsibility of scientific water control. The complexity of managing and protecting the water environment requires that water management be based on professional knowledge, so river basin supervisors must assume professional responsibility for water management. In practice, local leaders who act as river supervisors involve professional forces inside and outside the system to carry out professional water management and ensure effective water management. Therefore, professional responsibility is the basic accountability mechanism for local officials in water management.

China has built a comprehensive accountability system that includes both bottom-up social monitoring, parallel independent institutional monitoring and top-down system accountability, with multi-level, three-dimensional and multi-faceted characteristics. Through the multiple monitoring channels of formal institutionalization and informal institutionalization, the public can provide bottom-up feedback on water control issues and influence higher levels of government power to promote monitoring effectiveness. Second, parallel and independent institutional monitoring. People's consultative conferences at all levels play a democratic oversight role in advising and managing river chiefs on existing water management problems, and audit bodies at the district level and above conduct audits of river chiefs' natural resources when they leave office, thereby fulfilling their oversight and accountability role and promoting the implementation of the people's public will for water management. The third option is top-down hierarchical oversight. In the hierarchical supervision of the river chief system, China operates according to the power system and strengthens the supervision of the People's Congress at the same level, the hierarchical supervision of the administrative body and the top-down supervision within the Party.

In connection with the establishment of the River Chief System, the construction of a three-dimensional accountability network, which includes five major accountability networks: internal Party accountability, administrative accountability, judicial accountability, People's Congress accountability and public accountability, conducive to preventing and resolving the problem of responsibility shortage in water management by local officials, urging local officials to conscientiously perform their functions and responsibilities in water management, and ensuring that local people's governments at all levels are responsible and accountable for defending the country. In the future, we should further improve the three-dimensional accountability network, consisting of intra-party accountability, administrative accountability, judicial accountability, People's Congress accountability and public accountability, to establish long-term mechanisms to guarantee the transition from the "River Chief System" to sustainable river governance.

2 From River Chief System to Future Sustainable River Governance

Prior to the introduction of the river chief system, functional departments were separate from each other, and each department had long been divided and governed in its own territory. The introduction of the river officer system provides an effective way to address the lack of authority, comprehensively solve the dilemma of "water control in Kowloon", strengthen information communication and cooperation among different departments, maximize the integration of forces from all levels and departments in decision-making, and realize horizontal and interdepartmental cooperative management. Through the establishment of the river chief office, the river chief system integrates all functional departments involved in water control, such as the Environmental Protection Bureau, the Housing and Rural Development Committee, the Water Bureau, the Agriculture Bureau, the Urban Management Committee, etc., to solve the problem of multiple management in river basin management through the river commissioner system, break the traditional division of labor at the departmental level, and achieve interdepartmental cooperation. In addition, other coordinated authorities have been established in various localities to improve the administrative integration of water control. For example, in Guangdong, the Guangdong Provincial Watershed Management Committee was established. To further promote interdepartmental coordinated water management, the Guangzhou Water System Construction Headquarters was established in Guangzhou, whose members include the heads of the construction committee, environmental protection bureau, planning bureau, municipal garden bureau and other units, as well as the mayors of various districts and counties and the mayors of county-level cities. These horizontal coordination mechanisms will avoid the disadvantages caused by chaotic management by multiple departments, through inter-organizational networks, using and coordinating the "number one" at all levels, leading management, joint consultation, full-time responsibility, etc., and establish a coordination mechanism to strengthen cooperation between departments, effectively integrate forces at all levels, and improve the fragmented status of public management in the river basin.

The management and protection of transboundary rivers and lakes are mostly organized and implemented by administrative regions at the provincial level. However, in the case of large rivers, there are inevitable contradictions between the regional fragmentation of management and overall management, system management and coordinated management of river basins, leading to problems such as uncoordinated upstream and downstream protection and development, and insufficient joint prevention and control at the provincial and departmental levels, which is not conducive to the overall protection of rivers and lakes with river basin characteristics and requires the coordination of river basin institutions with different administrative regions.

In implementing the responsibility of the river chief system, the river chiefs are the decisive force. However, grassroots river chiefs face problems such as imbalance of power and responsibility, lack of capacity, and formalism, making the implementation of the river chief system a "last mile" problem. The implementation of the river chief system has reshaped the balance of power and responsibility in local water management, creating a relationship of river chiefs' accountability with clear responsibilities and strong commands. However, in the power distribution, there is still some imbalance between the powers and responsibilities of local river chiefs, especially the problems of unequal powers and responsibilities of grassroots river chiefs such as "small powers, many things", "small powers and large responsibilities" and "weak resources" are still more prominent.

The unequal powers and responsibilities in local water management are first reflected in the fact that there is less power and more work in the distribution of powers, which is particularly evident among grassroots river chiefs. For example, river supervisors in cities and villages are responsible for a large number of specific tasks in river monitoring and water control, but they do not have corresponding management powers, so in some areas of water control there is a phenomenon of "responsibility without power," which makes it difficult to solve some problems of water management at the grassroots level. At the same time, in the practice of water control, there is a certain conflict between the allocation of power for homogeneous responsibilities and the functional logic of territorial management, and the requirement of upward centralization puts grassroots river chiefs in a weak position in the allocation of rights and responsibilities, and even assumes many unnecessary tasks, which practically increases the burden of grassroots river chiefs in water management.

The unequal power and responsibility in local water control is also reflected in the low power and responsibility of grassroots river chiefs in the distribution of power. Due to the lack of scientific assessment and accountability, such as the phenomenon of "one size fits all" in the establishment of indicators, the frequency of assessment and accountability is inadequate, and some assessment and accountability are mere formalities, resulting in the alienation of river chiefs' assessment and accountability for water management in some regions. For instance, in 2018, the central environmental protection inspector found that the new music section of Shijiazhuang in Dashahe received full marks in the "river patrol" assessment despite many problems. In addition, in some areas, there is a phenomenon of shifting the responsibility of the river chiefs downward and focusing mainly on the river chiefs of cities and villages, such as the unannounced on-site visits of the river chiefs at the two levels of Guang'an City and Wusheng County in 2020, which found that the pollution of the Changtansi section of Sanxi City was severe. Due to the downward shift of the accountability target, some regions have achieved the phenomenon of evading responsibility through accountability, which reduces the effectiveness of monitoring and accountability of assessment and accountability.

The effective alignment of management resources is not only an important part of the allocation of rights and responsibilities, but also a strong guarantee of long-term water control. However, grassroots river chiefs face the phenomenon of multiple affairs and scarce resources, which affects the realization of the goal of transforming the "river chief system" into "river chief governance". Under the pressure-based system driven by the political tasks of ecological water control and the pressure of evaluation and accountability, although grassroots river chiefs can temporarily concentrate manpower, material resources and financial resources to fight for water control and achieve certain water control results, they still fail to change their practical dilemma, such as. Such as the lack of talented teams, insufficient support funds, and the shortage of professional equipment, which has a negative impact on long-term local water control.

The establishment of the River Chief System has strengthened the professional responsibility of local officials for scientific water management and created a new situation for water management in the new era. However, under the requirements of scientific water management, grassroots river supervisors still have significant deficiencies in their entrepreneurial ability, which affects the thorough promotion of water management in the new era. In the practice of local water management, most river chiefs are simultaneously appointed by local party and government leaders at all levels, and some river chiefs still have cross-appointments under multiple water management requirements, which leads to an overload of the role of river chiefs in water management, resulting in local chiefs lacking water management skills. "River chiefs" are not directly involved in the management of local water management, nor are they experts and scholars in the field of water management, which leads them to perform their roles in water management rather superficially. Due to the heavy workload, many tasks ended up falling on the head of the river commissioner's office, which limited the improvement of the river commissioner's operational capabilities for water management.

Due to the high professionalism of river management, it is urgent to build a high-quality and professional water control team in the local water control practice. However, in practice, the quality of grassroots river chiefs' water management teams still needs to be improved. As an important coordinating and liaison organization for water supervision, the staff of the river chief office should have relatively solid expertise in water management. In practice, however, River Chief Office staff are mostly temporary workers who have been transferred from other units, which not only makes it difficult to ensure work schedules but also leads to a lack of expertise. Due to a lack of expertise and operational skills in scientific water monitoring, some river masters at the grassroots river chiefs can only perform relatively simple work such as river patrols and flotsam recovery, while they cannot do their best in highly specialized water management matters.

In the practice of local water control, it is also necessary to apply the information management platform and technical scientific and technological means, such as various river patrol APP, problem monitoring APP, mass political platform, data sharing platform, the use of river patrol drones, etc., all of which require the mastery of certain technical application skills. However, grassroots river chiefs in some areas, grassroots river guards are older, and in some places environmental protection volunteers have acted as civilian river guards with relatively rudimentary scientific and technological application skills, making it difficult to adapt to the requirements of the new era of "Internet + water governance", which reduces the practical effect of scientific water control.

The grassroots river chiefs are the key figures in implementing the river chiefs' responsibility for water control. However, in the practice of local water control, some grassroots river chiefs have caused deviations in the implementation of water control responsibility, such as selective implementation, symbolic implementation and covert implementation and formalistic hidden concerns due to multiple considerations, and even the problem of "last mile" in the implementation of river chiefs' water control responsibility. Although they are at the end of the implementation chain of the river chief system, river masters at the grassroots river chiefs such as in communities and villages, are the final executors and supervisors of systematic corrections such as water resource protection, interception of wastewater, bank stabilization, and dredging. Since they are at the forefront of water control, they have sufficient information on water control, have some discretion in the practice of water control, and can escape supervision through their own information advantage. The grassroots party and government leaders are faced with a variety of complex tasks, and water management is only one of them, and it is very urgent only at a certain stage (such as inspections).

With the intensive promotion of the river chief system, the reform and exploration of "internet + water governance" is increasing, and "river chief APP", "river chief" and "river patrol" are emerging one after another, providing strong support for the refined management of rivers and lakes. However, in the context of the widespread application of technical governance, a symbolic implementation of formalistic water control has also emerged. To meet the frequent inspections and acceptance by superiors, grassroots river masters use the characteristics of Internet memory to write articles about "leaving traces" often enough, and even the problem of coping with formalism with formalism has occurred, such as. For example, the alienation of some river masters from the patrol action of discovering problems to a river patrol control, which is "to go around the river, take a look in the river, give a slap on the cell phone and pass the platform".

Due to the weak sense of responsibility, and imperfect fault tolerance and correction mechanism, some grassroots river police officers preferred to conceal the implementation of water control responsibilities. According to the requirements of the system, grassroots river chiefs find problems that are difficult to solve in daily monitoring, such as river patrol, and must report them to higher units in a timely manner through software platforms, such as the river patrol APP, but due to the principle of "territorial management", this may not only cause its own feedback problems, but will be reassigned to the local area, and the more complex and difficult the problems to be reported, the greater the likelihood of their own accountability. Out of a consideration to avoid liability risks, grassroots river chiefs are either unwilling to provide feedback on problems, or are evasive in reporting problems, attempt to change data, or are evasive and delay processing problems, resulting in incomplete and unobjective information feedback.

The grassroots river chiefs are the key figures in implementing the river chiefs' responsibility for water control. However, in the practice of local water control, some grassroots river chiefs have caused deviations in the implementation of water control responsibility, such as selective implementation, symbolic implementation and covert implementation and formalistic hidden concerns due to multiple considerations, and even the problem of "last mile" in the implementation of river chiefs' water control responsibility. Although they are at the end of the implementation chain of the river chief system, river masters at the grassroots river chiefs such as in communities and villages, are the final executors and supervisors of systematic corrections such as water resource protection, interception of wastewater, bank stabilization, and dredging. Since they are at the forefront of water control, they have sufficient information on water control, have some discretion in the practice of water control, and can escape supervision through their own information advantage. The grassroots party and government leaders are faced with a variety of complex tasks, and water management is only one of them, and it is very urgent only at a certain stage (such as inspections).

With the intensive implementation of the River Chief System, the reform and exploration of "Internet + water governance" is increasing, and "river chief APP", "river chief" and "river patrol" are emerging one after another, providing strong support for the refined management of rivers and lakes. However, in the context of the widespread application of technical governance, a symbolic implementation of formalistic water control has also emerged. To meet the frequent inspections and acceptance by superiors, grassroots river masters use the characteristics of Internet memory to write articles about "leaving traces" often enough, and even the problem of coping with formalism with formalism has occurred, such as. For example, the alienation of some river masters from the patrol action of discovering problems to a river patrol control, which is to "go around the river, take a look in the river, give a slap on the cell phone and pass the platform".

Promoting public participation in water management and forming a benign pattern of coordinated multi-subject water management is an important guarantee for promoting the "river chief system" and realizing the "river chief governance". However, in the current operation of the river chief system, public participation is still insufficient in many respects. The top-level public participation design is inadequate, the level of public participation needs to be improved, and the normal participation channels are inadequate. From the perspective of top-level design, public participation documents lack a comprehensive view, so public participation in the top-level design process is inherently inadequate. This is most evident in the following aspects.

First, public participation faces the dilemma that the legal system is not yet perfect. Although the public participation related to the establishment of the river chief system is politically supported by the "Opinions on the Comprehensive Implementation of the River Chief System", the stage of participation, the scope of participation, the participants, the legal implications and the limitations related to public participation are not clearly defined at the national legal level, and no legal limits are set on the scope, time and efficiency of participation, so that public participation in water management faces the practical problem of insufficient support from the higher-level law. From the perspective of the overall environmental protection legal system, most of the relevant public participation laws in China are abstract and principled provisions that lack clear and specific system design, are impractical, and are not effective enough to guide practice.

Second, there is the problem that public participation in policy documents is not systematic. The provisions on public participation in the "Opinions on the Comprehensive Implementation of the River Chief System" are also not perfect, as they only establish the requirements for expanding the channels for public participation and improving the awareness of public participation, but do not include public participation in the whole process of river chief system operation, and the provisions on public participation in decision-making, supervision and management are still relatively vague, so that public participation in water management does not provide perfect institutional guarantees.

Third, there is a lack of systematic provisions for exploring local public participation practices. In formulating the implementation plan for the River Chief System although some provinces emphasize the role of public participation in the operation and management of the river chief system, they are still limited to general calls and emphasize public participation as part of the operation and management of the river chief system, and there are few specific regulations and systematic explanations on public participation in water management, which limits the practical development of public participation in water management.

Currently, public participation in water management is not yet perfect, and the frequency, breadth, and depth of public participation in water management needs to be further deepened and improved. This is mainly reflected in the following aspects.

First, the frequency of public participation in water management is not high enough. Due to the limited awareness of public participation, public participation in water management is still not frequent enough, and participation in water management has not been integrated into daily life and behavior habits. Most of the existing forms of public participation are carried out by government organizations or municipalities, and the forms of participation lack systematicity and continuity, making it difficult for the public to fully express their interests in daily participation practices. For example, in the field of water management, authorities organize hearings and meetings to obtain the public's opinion, but these participation practices are often characterized by a certain degree of arbitrariness, and it is difficult to ensure normal public participation.

Second, the scope of public participation in water management is relatively limited. From the perspective of stakeholders, at present, participants in local water control are still limited to certain groups such as environmental volunteers, government agencies and school personnel, and public participation in water control still has the problem of too narrow participation. With the enhancement of the public's awareness of environmental protection, the public's enthusiasm to participate in water control has greatly improved, but it is still inconsistent with the expectations of public participation in diversified water control, and the initiative and sustainability of public participation in water control need to be further cultivated. At the same time, the environmental ethical responsibility of enterprises urgently needs to be strengthened. Due to vested interests and other factors, enterprises have limited motivation to participate in water monitoring, and their understanding of their own environmental ethical responsibility is vague, so that they even become polluters of water pollution. The enthusiasm of environmental protection organizations to participate in water control is very high, but they lack a systematic scientific understanding of water environment control, which can easily lead to misinterpretation and exaggeration of environmental problems and even have negative effects.

Third, public participation in water governance needs to be expanded. From the perspective of participation, the main limitation of public participation in water management is that individual participation in some areas is only superficial and does not play a substantial role, and the depth of participation needs to be expanded. For example, in the public opinion gathering stage of environmental quality assessment, in some areas, often only notices are published, and no one explicitly speaks out against it, which not only cannot reflect the true will of the people, but also makes it difficult to strongly support scientific and democratic decision-making. From a participation perspective, public participation is mainly focused on monitoring and is not integrated into the overall process of operating and managing the river basin system.

Normalized participation channels are an important guarantee for effective public participation in water management. At present, there are no normalized participation channels in China, which affects the public's enthusiasm and feeling to participate effectively in water control.

First, the traditional participation channels rooted in tradition need to be regulated. In the practice of public participation in water control, the existing channels of public participation can still have institutional effects, such as various kinds of consultation meetings, hearings, survey meetings, etc. However, the existing conventional participation channels still have shortcomings, such as low degree of institutionalization, great arbitrariness in making changes, and insufficient responsiveness, such as the fact that the implementation plan of local river supervisors for water regulation is mostly obtained within the government system, public participation and publicity are insufficient, mass participation is insufficient, and government agencies do not sufficiently respond to the opinions and suggestions of the masses. Based on the consideration of making full use of conventional participation channels, this should be improved through innovative working mechanisms, institutional procedures, and innovative outreach and mobilization.

Second, the use of innovative technology platforms is insufficient. In the information age, the increased use of information technology platforms for public participation in water management is of great benefit to improving the impact of public participation in water control. In the practice of local water control, the emphasis on "Internet + water control" has made technical means become more and more important support for river chiefs to control water, and various river patrol apps and information sharing platforms have emerged one after another. However, the information platforms are mostly limited to the internal use of government departments and officials, and their openness to the public is limited, and although the online consultation platform for the public to participate in water control has been explored, there is still much room for improvement.

Third, the opening of participation channels based on practice is insufficient. In the practice of local water management, it is particularly important to create innovative participation channels that are appropriate to the characteristics of water management. However, in some regions, insufficient attention is paid to this and the system is inadequately designed, resulting in some fragmentation of practice. For example, there is a lack of systematic planning and innovative channels for regular participation of civil river chiefs in water management, communication and cooperation with official river chiefs, and coordinated water management between civil river chiefs and official river chiefs, an important issue in water management. Although channels for exploring and participating in the design of cooperation mechanisms between civilian and official river chiefs exist in some regions, they have not been promoted in practice as a stable institutional approach.

Under the current hierarchical management model, the professional division of labor structure of management work is in conflict with the natural integral characteristics of the river basin, which is an important reason for the fragmented pattern of water environment management. As the fragmentation of departments and divisions becomes more and more difficult to meet the requirements of integration, functional diversity and benefits of water resources management in river basins, China has introduced a system of river directors, which aims to solve problems such as conflicts between departments and functions through effective hierarchical coordination and improve the synergy of water management. Through evaluation, the director integrates the responsibilities of items and blocks at each level. In hierarchical coordination, the system of responsibility of the chief executive relies on the administrative authority of the higher level, strengthens the responsibility of the chief executive at all levels through supervision and evaluation, and integrates the resources of different departments at each level through special agencies.

Specifically, this emphasized the establishment of a system of accountability in evaluation. Local party and government leaders are responsible for the overall management of each water supervision department and bear responsibility for water management in each jurisdiction as a whole, and through the layer-by-layer evaluation method, the task performance and management effectiveness of river managers at all levels are evaluated annually, and those who fail the evaluation are questioned, criticized and held accountable, etc., and some municipalities have also adopted a "one-vote veto system". The second point is to establish a monitoring system. In order to solve the problem of information asymmetry and improve the real understanding of lower-level governance, higher-level governments, including the central government, have established an inspection system. Through top-down inspection, it has exerted a certain deterrent effect on local parties and governments, and effectively promoted the fulfillment of local governments' responsibilities. In addition, some local governments have established reward systems, reward replacement systems, ecological compensation systems and bonding systems to motivate local government officials for water management. In general, the river head system urges river heads at all levels to effectively integrate the resources of various functional departments through the evaluation and accountability of superiors to subordinates, so as to achieve intervention and control of governance subsystems at all levels.

Government agencies, political leaders and public servants at all levels, as agents, are either responsible to the people's Congresses (entrusted organs) at all levels or to the higher organs entrusted with their functions and powers. Due to various reasons, agency problems often arise in the principal-agent relationship. Accountability mechanisms have been developed to limit the exercise of discretionary power by the client and to ensure the fidelity of the agent to the client. The premise of accountability is that the agent provides the principal with necessary behavior information and explanation. If government spending requires the consent of citizens and due process, resources must be used not only effectively, but also to achieve the desired results. The sustainability of effective transparency policies depends on three factors. First, the scope of targeted and transparent information should be continuously expanded around issues that are of public concern or need to be addressed, but under the premise of increasing the benefits to those who disclose information and those who receive information as much as possible, does not increase the cost of the information discloser and the information receiver. In order to protect the people's right to know, improve the transparency of government work and promote administration according to law, the regulations on the disclosure of Government of the People's Republic of China information came into force on May 1,2008. To further promote openness and transparency in government affairs, the General Office of the Communist Party of China and the State Council General Office issued and implemented the opinions on comprehensively promoting open government affairs on February 17, 2016, on November 15, the State Council General Office issued the detailed rules for the implementation of the opinions on comprehensively promoting the work of opening government affairs. Therefore, it is not only the basic requirement of the government's own work, but also the minimum requirement of the People's right to know.

In democratic politics, the subject of power must respond effectively to the interests of the members of the society (Archon, 2006). Political response has become an important link between the subjects of power and the members of society. When people have access to transparent information and engage in social accountability, they will expect government departments to respond to specific issues. Public sector responses to social accountability can be positive or negative. On the negative side, the public sector may ignore the reputational damage that could result from being held to account and even retaliate against specific accountability holders. In order to avoid a negative response from the public sector, China not only requires governments at all levels to actively promote public participation, but also to regulate the bottom line behavior of the government and local officials through legislation. For instance, the detailed rules for the implementation of the "opinions on the comprehensive promotion of open government affairs" issued by the State Council General Office clearly require that important decisions involving major public interests and public interests, we should not only expand public participation through various means, but also actively explore new models and build new platforms for public participation. In accordance with the requirements of the opinions on the full implementation of the river chief system, not only shall the performance of the river-leader of leading cadres be examined, but also shall the assets of natural resources of leading cadres be audited, even carrying out the system of life-long investigation of ecological environmental damage liability.

The implementation of any public policy not only needs the top-down power to promote, but also needs the bottom-up public participation. The People's willingness, initiative and active participation cannot be separated from the situation mobilization led by the government. During the course of the implementation of the river-long system, there were not only intensive propaganda reports by the central authoritative media, but also important speeches by the leaders of the party and the state as well as the leaders of the ministries and commissions, there are also local government propaganda departments through a variety of ways to inform the people of the implementation of the significance of the river system, far-reaching impact and specific practices, and so on, to create momentum for the imperative. The main reason why more and more people have been mobilized is that the propaganda mobilization has activated the consensus of the people on the river regime. The inherent logic of this consensus is that the government not only pointed out the urgency of water pollution control action, but also explained the effectiveness of the "one river (lake) one policy" action under the river system, it also inspires people to act responsibly and to expect the positive impact of their participation on sustainable water governance.

Reference

Archon, F. (2006). Varieties of participation in complex governance. *Public Administration Review, 66*(s1), 66–75. https://doi.org/10.1111/j.1540-6210.2006.00667.x

The manufacturer's authorised representative in the EU is Springer Nature Customer Service Centre GmbH, Europaplatz 3, 69115 Heidelberg, Germany. If you have any concerns regarding our products, please contact ProductSafety@springernature.com

Printed and bound by CPI Group (UK) Ltd, Croydon, CR0 4YY
30/09/2026
02225264-0006